Solar
Voltaic
Cells

ENERGY, POWER, AND ENVIRONMENT

A Series of Reference Books and Textbooks

Editor

PHILIP N. POWERS
Professor Emeritus of Nuclear Engineering
Purdue University
West Lafayette, Indiana

Consulting Editor
Energy Management and Conservation
PROFESSOR WILBUR MEIER, JR.
Head, School of Industrial Engineering
Purdue University
West Lafayette, Indiana

Additional Volumes in Preparation

Solar Voltaic Cells

W. D. Johnston, Jr.

Electronics Research Laboratory
Bell Telephone Laboratories, Inc.
Holmdel, New Jersey

MARCEL DEKKER, INC.   New York and Basel

Library of Congess Cataloging in Publication Data

Johnston, Wilbur Dexter.
  Solar voltaic cells.

  (Energy, power, and environment ; v. 7)
  Bibliography: p.
  Includes index.
  1. Solar batteries. 2. Photovoltaic power genera-
tion. I. Title. II. Series.
TK2960.J63   621.31'244   80-14158
ISBN 0-8247-6992-9

MARCEL DEKKER, INC.

270 Madison Avenue, New York, NY 10016

Current printing (last digit):

10 9 8 7 6 5 4 3 2 1

PRINTED IN THE UNITED STATES OF AMERICA

To Anne

Preface

A quarter century ago, in April of 1954, Daryl Chapin, Calvin Fuller and Gerald Pearson reported their invention of the silicon solar cell at Bell Laboratories *(Journal of Applied Physics, Vol. 25,* p. 676). Their first devices converted sunlight into electric power with an efficiency exceeding 6 percent, which represented more than an order-of-magnitude increase over the then extant state-of-the-art for direct solar-to-electrical power conversion. During 1955 a doubling of cell efficiencies was achieved and initial field trials in the Bell System demonstrated the technical and engineering practicality of silicon solar cells for power generation. The costs were very high, however, in comparison to traditional sources of electricity, and hence use of the cells was not economically justifiable.

The growth of the United States space program during the 1960s spurred the development of solar cell technology as a source of power for earth satellites and missions of space exploration, and a small but active manufacturing industry was formed. Prices remained much too high for general use on earth, however. The 1970s have produced a public awareness that nontraditional energy sources must be developed to supplement and eventually to supplant our near total present reliance on nonrenewable fossil fuels. An increased level of interest and activity in research and development in the photovoltaics field has ensued as a result, stimulated by greatly increased support from the United States government. Active research and development aimed at reducing the cost of terrestrial solar cell arrays and systems is now also underway in many other countries as well.

The purpose of this book is twofold: first, to present an up-to-date picture (i.e., as of March, 1980) of where we stand today on the road to a practical solar photovoltaic electric capability which could make a *significant* contribution to the energy needs of the United States and the world; and second, to provide in as nontechnical a way as possible an explanation of what solar cells are, how they work, and what lines of research are being pursued to improve present performance or reduce present costs (or both). A reader with some background in engineering or physical science at the undergraduate level has been assumed. The second chapter, for example, presents the more specialized concepts of solid-state and semiconductor physics needed to understand solar cells, building from an elementary background in modern physics. The discussion of new research directions in the fourth chapter is largely materials oriented and an understanding of inorganic chemistry and thermodynamics (or physical chemistry) would be of value. The fifth and sixth chapters contain discussion which includes economic aspects of photovoltaic systems. In looking toward the future, it is at best possible to delimit the scope of plausible scenarios and point out various factors which will be important and the probable ranges of those factors which will allow large-scale use of terrestrial solar-cell arrays to come to pass.

Much of the content of this book touches on uncertain ground: Will single crystal silicon solar cells every be truly inexpensive? Are CdS/Cu_2S solar cells capable of long life and good efficiency? Is there some better material? In spite of a conscious and deliberate effort to balance enthusiasm and skepticism, this book is, ultimately, an expression of my own opinions in these and other areas of current controversy. The number of untried ideas and the areas of unknown territory have been much reduced in the last year or two, however, and the probable directions for further development of practical terrestrial solar cell arrays have been greatly clarified as a result.

I very much appreciate the support and encouragement provided by the management of Bell Laboratories which allowed me to undertake writing this book. I have benefitted greatly from the atmosphere provided here by my colleagues on the technical staff. I am also grateful for many informal discussions with members of the growing photovoltaic community from other institutions, many of whom are cited formally in the text references. I would like to thank Mr. S. J. Bennett for supervising the production of the artwork, and Mrs. E. Smolinski, Mrs. C. King, and Mrs. A. C. Applegate for carrying out the many word processing details associated with preparation of the type set manuscript.

W. D. Johnston, Jr.

Contents

Chapter 5
BEYOND THE CELL   **149**

CHAPTER 1

Solar Energy and Energy Needs

The "energy crisis" Americans experienced in the Fall of 1973 and Winter of 1974 was not an *energy* crisis in actuality. The tripling of Near-Eastern oil prices at that time caused trade imbalances which continue to pose serious problems for the economic stability of industrialized nations in the free world. These events, together with the motor-fuel shortages of May-June, 1979 precipitated by the Iranian revolution, should serve to bring home the realization that a *true* energy crisis *is* building and cannot be avoided so long as the world continues to rely on fossil fuel sources, or indeed on any finite terrestrial source, for its energy needs. The *only* long-term solution lies in the utilization of solar energy for most, and eventually all, of mankind's energy requirement.

This book will address the principles and present level of development in practice of *one* approach to utilization of solar energy: the use of solid state devices which allow the direct conversion of sunlight to electricity. This approach has a substantial number of potential advantages, including the absence, in principle, of moving mechanical components subject to wear; the absence of radioactive waste or exhaust, other than waste heat; and the absence of a need to build large-scale, multigigawatt plants to achieve operating thresholds or optimum efficiencies. There are also limitations, of course, some of which are fundamental, and some of which reflect our present imperfect control of the materials processing and fabrication of photovoltaic devices. In common with other solar energy approaches, many of the limitations of photovoltaic energy conversion derive from the properties of terrestrial sunlight. In this chapter we will consider the match between sunlight as energy source and energy needs, focusing principally upon the United States.

1.1 Sunlight

Our sun is a rather ordinary main-sequence star with what is for practical purposes a constant radiative energy output, derived from nuclear fusion reactions. In every second about 6×10^{11} kgm of H_2 is converted to He, with a net mass loss of about 4×10^3 kgm, which is converted through the Einstein relation ($E=mc^2$) to 4×10^{20} J. This energy is emitted primarily as electromagnetic radiation in the ultraviolet to infrared and radio spectral regions. The total mass of the sun is now about 2×10^{30} kgm and a reasonably stable life in excess of 10^{10} years is projected.

The intensity of solar radiation in free space at the average distance of the earth from the sun is defined as the solar constant with value of 1.940 Langleys per minute (1.94 cal/cm²·min, 1.353 W/m², 428 BTU/ft²·hour). Because the earth moves in an elliptical orbit, the value of solar intensity at the upper atmosphere varies by about ±3 percent during the course of a year. Much more significant variation results on the earth's surface because of climatic factors and, of course, the day-night cycle due to the earth's rotation.

For thermal solar energy collection it is usually a good approximation to assume that the solar spectrum is that of a 5800K black body. The output of a solar cell, however, is significantly affected by the details of the spectrum of incident radiation and terrestrial sunlight deviates in important ways from a black body spectrum [1], as shown in Fig. 1.1. The principal effects are due to water vapor absorption in the infrared, ozone absorption in the ultra-violet, and scattering by airborne dust and aerosols. Aerosol scattering varies inversely with the fourth power of the wavelength (which is why the sky appears blue rather than yellow or red). This scattering causes a reduction in the blue content of *direct* sunlight and an increase in *diffuse* sunlight or *skylight*.

During the course of the day, the angle between the earth's surface and the sun varies. The effective area of a fixed solar cell array (that projected area normal to the sun's direction), thus changes. While this may be compensated with a *tracking* array, the position of which is continually adjusted so as to point at the sun, the path-length of the sun's

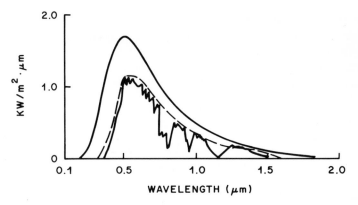

Figure 1.1 Spectral power distribution for a 5800K black body (upper solid curve); resulting spectrum after attenuation by ozone and aerosol scattering (dashed); actual AM2 spectrum (lower solid curve) showing water vapor absorption structure. (Adapted from Ref. 1).

light through the atmosphere will still vary. The secant of the angle between the sun and the zenith is called the *air mass*. It is a measure of atmospheric path length relative to the minimum path length when the sun is directly overhead. For large air mass values, the atmosphere has significant refractive effects but these are normally ignored since only a small portion of the daily total of solar energy is received under such conditions.

For purposes of calculation, or for comparison of research measurements made on different cells in different parts of the country, it is useful to define a standard terrestrial spectrum in terms of specific representative values for water vapor, ozone, and aerosol concentration in the atmosphere. Air mass 1.5 conditions (sun 45° above the horizon) represent a satisfactory energy-weighted average. The spectrum obtained by calculating the effect of water vapor, ozone and aerosol scattering on the measured extra-terrestrial spectrum [2] is shown in Fig. 1.2. The total energy flux for this spectrum is 935 W/m^2. In terms of round numbers, a value of 1 kW/m^2 is often taken as representative of peak daytime sunlight intensity at sea level and this intensity is referred to as one sun.

In addition to peak energy, it is also of interest to know how much solar energy can be expected over the course of a year in various locations. This *insolation* data has been gathered in several locations in the United States for a number of years, and much more data has recently become available as interest in solar energy has intensified. Not surprisingly, the areas known for sunshine have more solar energy

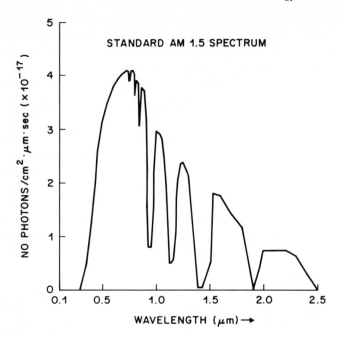

Figure 1.2 Standard (defined) solar flux spectrum for AM 1.5 conditions. (Data from Ref. 2).

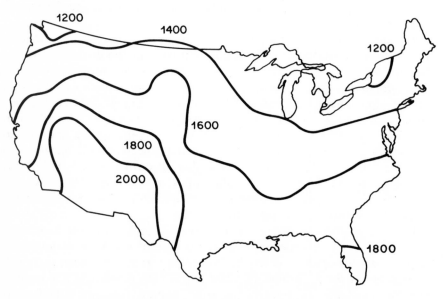

ANNUAL INSOLATION CONTOURS-KWhrs / m² · yr

Figure 1.3 Distribution of solar energy incident (horizontal surface) on an average annual basis for the United States. (Adapted from Ref. 3).

available over a year than those areas known for overcast weather. What may be surprising is that the *extremes* represent *only* about a factor of two variation. The annually averaged solar energy contours for the United States [3] are shown in Fig. 1.3.

1.2 United States Energy Requirements

The historical [4], and projected, trend of energy use in the United States is shown in Fig. 1.4. A number of scenarios have been used to predict future trends. All assume demand for energy, and use of energy, will grow, but it appears likely that the rate of growth must slow appreciably from that which has characterized the recent past. The traditional unit for measurement of energy demand is the quad (Q), equal to 10^{15} (a quadrillion) BTUs. Various useful energy equivalents are listed in Table 1.1. SI units (joules/m^2, etc.) are preferred but for some purposes the older units are particularly convenient and have remained in use.

Historically the energy consumed in the United States has come dominantly from one source at a time. First wood, then coal, and now liquid fossil fuels have dominated the energy supply. The projection shown in Fig. 1.5 becomes difficult to defend in detail beyond the year 2000, although continuation of present trends in nuclear plant construction and oil and gas finds seem likely. The particular projection plotted derives from U.S. government studies conducted in 1974; if anything

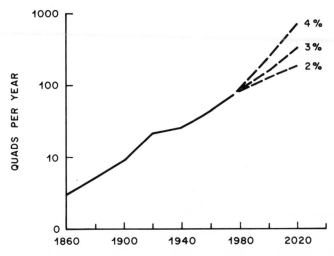

Figure 1.4   Growth of energy use in the United States, and projection assuming average annual rates of 4 percent (average post-1945), 3 percent (average for last century), and 2 percent (optimistic estimate including impact of conservation programs). (Data from Ref. 4).

Table 1.1
Energy Equivalents

	BTU	kWhr	bbl	ccf	tons
1 Quad	10^{15}	2.9×10^{11}	1.7×10^{8}	9.8×10^{9}	4×10^{7}
1 kWhr	3.4×10^{3}	1	5.8×10^{-4}	3.3×10^{-2}	1.4×10^{-4}
1 barrel of oil	5.8×10^{6}	1.7×10^{3}	1	5.7	0.23
100 cu ft nat gas	1×10^{5}	30	1.7×10^{-2}	1	4×10^{-3}
1 ton of coal	2.5×10^{7}	7.4×10^{3}	4.3	247	1
1 Acre-yr sunlight[a]	1.34×10^{11}	4×10^{7}	2.3×10^{4}	1.3×10^{6}	5.6×10^{3}

[a] Along the 1600 kWhr/m^{2}yr contour of Fig. 1.3. Data from Ref. 4.

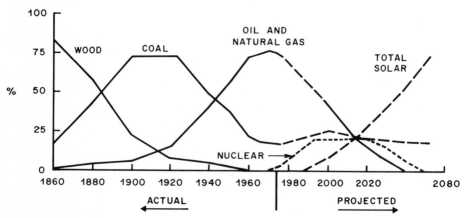

Figure 1.5 Historical sources of U.S. energy (data from Ref. 4) and a possible projection for the next century. To the extent that solar energy does not become available to meet this projection, and provided environmental restrictions on operation of coal-fired and nuclear electric generation are not relaxed, a growing energy shortfall will result.

the growth of nuclear energy exploitation has been slower than suggested here. Numerous other studies have been made; most agree only in that the only energy source making an increasing contribution by the mid 21st century is solar.

During the last two decades the percentage of total energy used in the United States in commercial/residential applications (excluding electricity) and for transportation purposes has remained nearly constant at 20 percent and 25 percent respectively. The percentage used in industry (excluding electricity) has *declined* from 36 percent to 25 percent over the last 25 years, while the percentage utilized for electric generation has nearly doubled going from about 15 percent in 1950 to 29 percent in 1976. The actual quantity of United States electric generation tripled in the 1960 to 1976 period, going from 0.75 to 2.04×10^{12} kWhrs. This production was met primarily from fossil fuel combustion — 46 percent coal, 16 percent oil, and 14 percent natural gas. Hydroelectric power constituted another 14 percent, nuclear fission 9 percent, and less than 1 percent was obtained from all other (mostly geothermal) sources.

On a world-wide scale, the United States energy consumption is dominant. About one-third of the world's energy is consumed in the United States by one-sixteenth of the world's population. The pressure for energy equity for less developed countries is increasing and it seems clear that long-term political stability will require that their needs be taken into account in setting the energy plans and policies in the developed industrial nations.

The development of solar energy is particularly appropriate and timely in this context since an abundant share of the earth's sunlight is the one resource undeveloped nations have in common. These countries do not have established central electric power distribution systems (grids), and there may not be the necessity to develop *totally* interconnected national electric systems. The dispersed nature of sunlight may thus be much less disadvantageous in underdeveloped countries than in industrial countries where power systems have already evolved so as to be compatible with centralized power sources geared to high energy density.

In the United States, however, the central electric power grid is a fact of life and any substantial solar electric contribution to United States energy requirements must be grid-compatible. Thus, if private individuals are to own their own solar electric generating equipment, appropriate systems must be developed to facilitate transfer of surplus power into the grid during peak sunshine hours and out of the grid during night-time demand. This raises the questions of utility buy-back rates and policies, as well as questions of equipment availability (and cost) to ensure mutual safety of the utility-owned and private equipment. These are not trivial matters, but if solar electricity is available on a practical basis there is little doubt they will be solved. The essential question is whether solar electricity can be generated at a cost which will permit its use without destroying an energy-intensive economy.

Electricity is produced from nonsolar sources by heat engines; i.e., gas (Brayton cycle) or steam (Rankine cycle) turbines, coupled to generators. Solar heat could, in principle, be substituted directly for the heat now obtained from fossil fuel combustion or nuclear fission. The traditional path for thermal electric generation has been "bigger is better". Efficiencies increase for large, high capacity plants which use large heat sources of high energy density. This produces an inherent mismatch, so far as solar thermal electric generation is concerned, with the dispersed nature of the solar resource. Photovoltaic electric generation, however, is subject to entirely different economies of scale in generator size. It is not necessary to construct multigigawatt plants since full modularity is achieved at very much lower levels, on the scale of kilowatts, in fact, as would be suitable for single-family use.

Photovoltaic conversion systems also offer freedom from moving parts and high temperature steam or gases, and hence should be reliable and require little maintenance. Environmental effects are minimal. No chemical exhaust is involved, no spent fuel or ash is produced (radioactive or otherwise), and no local thermal pollution is generated owing to the low energy density. The serious question is one of cost and performance: could photovoltaic systems be built in quantities that would permit impact on the 2×10^{12} kWhr annual usage scale?

It is easier to see the scale involved by considering the electric needs of an average family — about 750 kWhrs per month in a single family residence of 200 m^2 floor area. At the latitude of Washington, D.C. the *average* insolation is 100 kWhrs/m^2 per month, so about 75 m^2 of solar cells operating at 10 percent efficiency would suffice in principle to provide the needs for such a living group. This area is compatible with a south-facing roof area for such a residence of ~100 m^2. About one-third of the electric use in the United States lies in the residential/commercial sector, thus roof area constraints do not preclude photovoltaic generation of up to one-third of electric (10 percent of total) energy needs.

1.3 U.S. National Photovoltaic Program

There is a considerable body of engineering experience with photovoltaic generation derived from the United States program of space exploration, essentially all based on Si solar cells. The scientific and engineering feasibility of this technology has been adequately demonstrated. Solar photovoltaic power is now the preferred primary power choice for most unmanned and all manned space missions. Prior to 1975 only limited terrestrial applications for photovoltaics existed, often using arrays of cells which had failed to meet acceptance criteria for space projects. In that year the Energy Research and Development Administration (ERDA) of the United States Government implemented a national photovoltaic conversion plan as part of the National

Solar Energy program. This action had been mandated by the United States Congress in October, 1974 (PL 93-473), with the stated objective of developing a "viable industrial and commercial capability to produce and distribute [photovoltaic] systems for widespread use in residential and commercial applications." The function of the ERDA has now been taken into the Department of Energy and a new (1978) National Photovoltaic Program Plan [5] has been announced.

The Solar Photovoltaic Energy Research Development and Demonstration Act passed in 1978 (PL 95-590) emphasizes the development and demonstration of commercially competitive technology with the goal of obtaining "electricity from photovoltaic systems cost competitive with utility-generated electricity from conventional sources". The latest revisions to the plan [6] specifically recognize the need for cost reduction in all system elements, although the major research emphasis remains on photovoltaic materials and solar cell technology.

The current program represents a refinement rather than a redefinition of the 1975 Plan. In addition to questions of efficiency and manufacturing cost, questions of system life, cost-over-life, marketing strategy, maintenance and reliability are being addressed. Basic research and advanced development funds are programmed for evaluation of photovoltaic technologies based on other materials than Si. Periodic market purchases of photovoltaic arrays are intended to stimulate the nascent industry now existing and to provide equipment for demonstration and evaluation. The scale of this program is large, with well over a billion dollars expenditure projected over the 1978 — 1988 decade.

The objectives are ambitious:

1. By 1982: Reduction in price of flat arrays of Si cells to $2. per peak watt (one sun) output, with an annual production capability of 2×10^7 watts peak. (The current figure is ~$9.50, down from $40. in 1975.)

2. By 1986: Price of $0.50 per peak watt at 5×10^8 peak watts per year production.

3. By 2000: Price down to $0.10 to $0.30 per peak watt at 10^{10} peak watts per year. This level should provide electricity at a cost-over-life equivalent to the $0.04 to $0.06 per kWhr cost projected for utilities 20 years hence.

All the above costs are given in 1975 dollars to take out the effect of overall inflation. The trend over the last several years and projected into the future [7] is shown in Fig. 1.6. In addition to the first cost

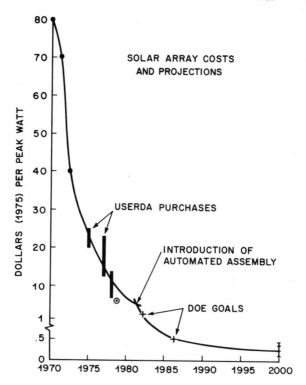

Figure 1.6   Actual purchase costs of silicon flat-plate arrays (bars) and a concentrator array (circle) in relation to DOE projection curve and goals. Data from Reference 5.

figures cited, an array lifetime of 20 years and a generation efficiency exceeding 10 percent are specified in the program goals. These figures are not really arbitrary. A photovoltaic electric system will *necessarily* be capital intensive (high first cost) since the fuel is free. The cost-over-life will then be strongly dependent on financing terms. A 20-year lifetime would appear the minimum to warrant long-term, low-rate borrowing. It can also be shown that there is a minimum array efficiency below which no net return on investment is possible even if the array itself is free. Enough electricity must be produced to cover the other fixed costs: installation labor, maintenance, siting expense, power conditioning electronics, etc.

Several studies have concluded that for many of the proposed applications with greatest potential impact a 10 percent array efficiency provides a minimum margin. Higher array efficiency is, of course, desirable but has a rather small (sub-linear) effect on overall system costs. For reasons we will examine in detail in the next chapter, 20 percent efficiency represents about the ultimate that can be expected theoretically for the kind of flat plate array we have been considering. Other schemes, using spectral separation and/or concentration of the sunlight by special optical devices, offer somewhat higher efficiencies at the expense of added complexity. It appears at this time that an array concept featuring minimum *acceptable* efficiency over a longer life (to 30 years, perhaps) with minimal or no maintenance would hold the edge over a less robust, higher performance array. This is particularly likely to be the case in residential installations, where skilled maintenance as a practical matter will not be available.

The stated role of the Department of Energy in stimulating the development of a large-scale photovoltaic industry will consist in funding support for technology development, for research and advanced process development, for systems studies, and for quality assurance and standards definition. Demonstration, system tests and public display applications represent roughly half the total budget. These are intended to demonstrate actual performance, familiarize the public with the realities of photovoltaics, and through purchase programs, encourage the growth of large volume, automated manufacture of solar cell arrays by private industry. The overall goal is, of course, to create market demand by making photovoltaic power generation economically and environmentally *the most attractive* energy option for many purposes in most of the country.

1.4 Economic and Social Implications

If the DOE plan is met, in the year 2000, the manufacture of solar photovoltaic arrays would be a billion-dollar industry. The eventual size would depend on growth of energy needs and escalation in price of alternative energy source fuels. The 1975 ERDA plan projected a 5 to 15 billion dollar sales volume for solar cell arrays by the year 2000. In either case, it would appear that if this program is successful a major new manufacturing industry must be established, on a scale approaching that of the present automobile industry. In terms of area, a production volume in the range of 0.1 to 1 billion square meters per year is envisaged. Clearly the analogy with the automobile industry is appropriate, as no other *finished* product except automobiles is manufactured in such quantities in the world today (not counting, of course, building materials like plywood, steel sheet, or glass).

One can, in fact, define an equivalent area associated with an automobile of perhaps a hundred square meters. This suggests that mass production of solar cell arrays at a cost of $0.50 per peak watt is not out of the question; the cost is then ∼$50. per m^2 for either an average automobile or solar array. The point of this analogy is that manufacture of a product *with some technological complexity* at the approximate cost and in the approximate quantities under discussion for the photovoltaics program *can be done*. Evidently similar materials (steel and glass), and similar methods (rolling, stamping, spraying and use of automated machinery for welding and fastening), must be employed.

There are other miscellaneous considerations which need to be mentioned to round out the picture, although they will not be considered in detail here. The solar component of property rights will have to be defined legally. The nature of the customer/supplier relation between property owners and central electric utilities will require redefinition. The effect of buy-back rates on the cost effectiveness and storage requirements of photovoltaic installations will be critical. The development of central disconnect control or other protection mechanisms for utility maintenance workers, the impact of surplus photovoltaic electricity on peak load pricing, and the need to maintain sunless day peak capacity, are other aspects which involve possibly difficult social technology and a restructuring of the accounting practices and policies of the regulated utilities.

A final challenge any energy solution must pass (particularly one enjoying or apparently requiring initial government subsidy and thus suspect of being unnatural in a free enterprise society) is that it be a *true* energy breeder. The useful operating life of a photovoltaic system must be long enough that *at least* enough electric energy is made available to produce its replacement system, *plus* enough for the useful output required. Fossil fuel systems are *not* true breeders, since they do not produce fuel, but they do provide surplus useful energy beyond that required for fuel gathering and power plant construction. On that basis, controversy has arisen occasionally as to whether net energy is generated or consumed by the world's nuclear electric reactors. Even the misnamed breeder reactor only converts otherwise nonfissionable U^{238} to fissionable material. It does not *create* new fuel but converts an existing, *finite* terrestrial fuel source to a usable (and expendable) form.

This is not the place for an analysis of nuclear reactor cycles, designs, fuel mining, extraction and concentration methodologies. Whether some net energy is made available or is actually lost overall in conventional or breeder reactor operation is not so significant as the mere fact that the point has been and is occasionally still seriously debated. Under these circumstances, it is clear that the nuclear energy balance has been much closer than one would desire. Careful attention to this problem in the case of photovoltaics is certainly necessary.

Early analyses of the energy pay-back time for silicon solar cells as manufactured for the United States space program suggested life-times in excess of *40 years* would be required, and hence such cells would not be *energy-cost-effective.* Those cells also cost at least 200 times too much to be considered *dollar-cost-effective.* More recent studies [8] based on *current* commercial technology for Si cells in flat plate, ready-to-install arrays intended for terrestrial application indicate a four year pay-back time for all the manufacturing energy including lighting and heating or air conditioning the plant buildings. The DOE has contracted with a commercial manufacturer of photovoltaic systems to construct and operate a solar breeder plant which will manufacture solar cells using *only* the solar electric and thermal energy input available from a hybrid thermal photovoltaic collector array. This will provide direct verification of the breeder concept, and an empirical demonstration of the energy pay-back time under at least one specific set of operating conditions.

1.5 Less-Developed Countries

The discussions in the preceding sections of this chapter have been centered on the energy needs and resources of the United States. The energy crisis, when it comes, will be worldwide and for that reason alone some comment on energy consumption and supply in the rest of the world is in order. The patterns of energy use in the industrial nations of Western Europe and Japan are not unlike those in the United States, but there is a difference of degree in present cost of energy and energy consumption per capita or per unit of gross national product. Energy is more available, cheaper, and is used more extensively (and probably less efficiently) in the United States than in the other industrial countries of the world.

Reference to the world insolation contours [9] in Fig. 1.7 indicates that the United States is favored slightly with solar energy resources relative to these other industrial nations. There are regions in the American Southwest which receive as much solar energy per year, on average, as is incident anywhere on the globe, and twice as much as is available in the northern European countries. The bulk of the inhabited regions of the solar belt between 40 ° north and south latitude fall within countries in which industrial development has lagged far behind that in the advanced countries, or has barely even begun. Nevertheless, a very substantial portion of the world's peoples live in this belt, including residents of both the richest and poorest countries on earth.

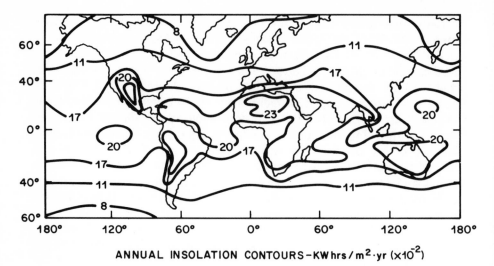

ANNUAL INSOLATION CONTOURS–KW hrs / m^2·yr (x10^{-2})

Figure 1.7   Annual average incidence of solar energy worldwide [8].

These countries may represent a substantial intermediate market for photovoltaic arrays. A study of the potential for photovoltaic irrigation water pumping systems in Chad, India, Bangladesh and Pakistan, for example, shows that solar electric systems become the least cost option when the cost of diesel fuel exceeds about $0.30 per liter and array costs between $1.30 and $0.50 per peak watt are realized [10].

Traditional energy studies have implicitly taken the United States model of rural electrification for granted; i.e., the assumption has been made that extension of a centralized power grid would take place. Such studies have then examined the various current options (fossil fuel, nuclear, hydroelectric) for powering the grid. The reality in less developed countries is that central power transmission grids are incomplete or even nonexistent, and that much or most of the current energy used is generated locally from renewable resources with a large input of human effort.

Centralized power generation and distribution is readily encompassed in national or regional political programs, but channeling funds for local development of decentralized power is much more difficult. Poor, rural villages are not uniformly inhabited by poor people of equal poverty level; there are social and political strata at the village level. In the several hundred thousand rural Indian villages of fewer than 500 inhabitants it is quite typical that fewer than 10 percent of the residents control more than 75 percent of the land and resources. When modern improvements of whatever kind come, the local elite tend naturally to be the first to benefit. It is unlikely that a technology switch such as the introduction of photovoltaic electricity will change this social reality.

From the point of view of developing countries, availability of photovoltaic power generation systems would be a mixed blessing. Solar cells are a high technology product and could not be produced in a cottage industry setting; arrays would have to be imported from the developed countries at a high cost in foreign exchange. The entire cost (or nearly so) of photovoltaic systems is *first cost* since the fuel is free, and requirement for very little local support labor is generated. This places special burdens on countries with limited capital and little or no foreign exchange resources.

Against these social and economic considerations are to be weighed the potential technological advantages of photovoltaic generation: reliability, modularity, independence of fuel supply, and freedom from pollution. Even if these advantages become available at competitive or even lowest *over-life* prices photovoltaic power systems may not be widely adapted in developing countries until *first-cost* prices become competitive as well. In spite of the obvious match between supply and presumed need, rapid acceptance of solar photovoltaic technology in developing countries does not appear assured.

1.6 Summary

The United States and the world in general derived the bulk of its energy needs from renewable, solar derived sources (wood) until the middle nineteenth century. Fossil fuels — first coal, and starting in the early 20th century, oil and natural gas — then became dominant. Consumption of the last has peaked and is now declining. Projection to the year 2000 indicates that even with substantial conservation measures serious energy shortfalls will be incurred relative to traditional domestic sources, *including nuclear fission* as utilized in current reactors. Imported energy will no longer be reliably available, either because of depletion, or because of global politics, as demonstrated plainly by the upheavals in Iran during 1979.

Solar energy is abundant in adequate measure to provide for the United States' energy needs, provided an adequate technology to harness it is developed. A realistic suggestion is that 25 percent of the nation's energy needs may be derived from all solar sources within 50 years. Perhaps one-third of the total electric needs may be met through direct photovoltaic conversion. For this to occur a large scale manufacturing industry will need to be established, capable of producing 10^8 to 10^9 m^2 per year of environmentally sealed solar cell arrays. These should cost $\sim\$30/m^2$ in 1975 dollars, be capable of 10 percent solar to electric power conversion efficiency, and last 20 years. Cost production targets on the path to this goal are $\$200/m^2$ and $10^6 m^2/year$ by 1982, $\$50/m^2$ and $10^8 m^2/yr$ by 1986.

These cost/performance goals can only be met through sophisticated high volume mass production using abundant and inexpensive materials. The capability to hold down costs of manufacture is critical and is the major technological challenge lying in the path of large scale photovoltaic power generation. Nevertheless, actual progress toward these milestones since 1974 has proceeded ahead of the forecast schedule [7].

The achievement of these cost-performance goals will not of itself guarantee acceptance of photovoltaics. The high first cost and extreme capital-intensive nature of any free-fuel power system is a deterrent which will favor other initially cheaper systems that actually cost more over the long run if present-day financing, investment, and tax depreciation criteria continue to be applied. This is a particularly difficult problem for the capital-poor developing countries of the world. Decentralized electric generation and use of solar energy systems in general appears more compatible with local cooperative/commune government than traditional Western centralized capitalism. With continually rising costs of alternative energy sources over the present day, a time will inevitably come when photovoltaic conversion is the least cost alternative on a first-cost or short term basis, and economics will then dictate that whatever social changes need to be made will occur. We may hope this time will occur sooner, through reduction of array costs; and energy costs will stabilize at an affordable level worldwide, without major social upheaval.

REFERENCES

1. M. P. Thekaekara, *Solar Energy* **14,** 109 (1973).

2. *Report# ERDA/NASA/1022-77/16 (NASA TM-73702)* (United States Nat. Tech. Information Service, Springfield, VA, 1977).

3. cf. J. A. Merrigan, *Sunlight to Electricity* (MIT Press, Cambridge, MA., 1975).

4. United States Department of Energy Publication *DOE/OPA-0020* (United States Government Printing Office, Washington, DC, 1978).

5. United States Department of Energy Publication *DOE/ET0035(78), UC-63* (United States Government Printing Office, Washington, DC, 1978).

6. United States Department of Energy Draft Document *DOE/ET-0105-D* (June 6, 1978, unpublished).

7. P. D. Maycock, *Proc. 13 PVSC* , 6 (IEEE, New York, NY, 1978).

8. M. Wihl and A. Scheinine, *Proc. 13 PVSC,* 908 (IEEE, New York, NY, 1978).

9. Adapted from Solar Energy Sales Brochure of Amperex Electronic Corporation (North Amer. Phillips), Slatersville, RI, printed 1978.

10. D. V. Smith, *Photovoltaic Power in Less Developed Countries* (Report #C00-49094-1), MIT Lincoln Laboratory, 1977 (unpublished).

CHAPTER 2

Basic Principles

As a practical matter, solar cells with usefully high efficiency to be of interest in engineering applications are fabricated from semiconductors. In this chapter we will discuss the electronic structure of semiconductors and the physics of electric current generation and collection in solid-state semiconductor solar cells.

2.1 Crystalline Solids [1]

When atoms are assembled into a crystalline solid, the sharp electronic energy levels characteristic of the individual atom merge. The valence electrons, particularly, are subject to interaction with the surrounding atomic nuclei and electrons. We can consider the crystal to be formed from an array of individual atoms having the same relative spatial arrangement as in the actual crystal but with a greatly magnified distance between atoms, and then conceptually let this distance be reduced to the actual crystalline interatomic distance.

The electronic energy levels of the system of widely separated atoms will be just those of the individual atom multiplied by a degeneracy factor equal to the number of atoms. This is a very large number, $\sim 10^{22}$ for real crystals of cm dimension. As the interatomic distance is reduced, this degeneracy must be removed and the sharp level splits into distinct levels. For the high degree of degeneracy involved it is permissible to consider these split levels as forming a *continuum*, since the spacing between levels will be of the order of 10^{-22} eV, much smaller than thermal broadening at any reasonable temperature ($kT \sim 10^{-4}$ eV/K). As the interatomic distance is further decreased, the relative strength of initial interactions increases and leads to increased broadening and eventually to overlap of the bands arising from different atomic levels. This is shown schematically in Fig. 2.1.

At the equilibrium position of the atoms we will have a series of bands which will, in the ground state, be filled to a certain point. The bands may overlap partially or be separated by greater or lesser gaps, depending on the nature of the atomic energy levels. The nature of electrical conductivity in a solid is governed primarily by the details of the *band structure* in the vicinity of the upper edge of the occupied states. It turns out that this same structure also determines the absorption (or transmission) of light, as well.

At a given nonzero temperature the electron states will be filled in accordance with the Fermi-Dirac distribution

$$n(\epsilon) = g(\exp((\epsilon-\mu)/kT) + 1)^{-1} \tag{2.1}$$

in which n gives the probability that a level of energy ϵ and degeneracy g is filled at temperature T, k is Boltzman's constant and μ is that energy (the Fermi energy) for which the occupation probability is 0.5. As $T \rightarrow 0$, $n(\epsilon) \rightarrow 1$ for all energies less than μ and is zero for energies greater than μ. The value of the Fermi energy at finite temperature depends on the density of states $\rho(\epsilon)$ (i.e., how the available states are

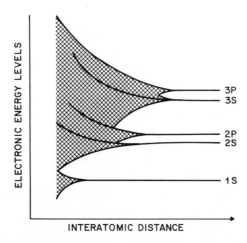

Figure 2.1   Electronic energy levels for a large number of regularly spaced atoms as a function of interatomic spacing.

distributed in energy) as well as on the temperature, and is determined from the normalization condition

$$N = \int n(\epsilon)\rho(\epsilon)\,d\epsilon \qquad (2.2)$$

with N the total number of electrons in the crystal.

A particularly simple illustrative case is that of alkali metals, which have a single s electron outside a closed p shell. In the alkali metal crystal the uppermost occupied s band will then be almost exactly half full, and these crystals will be metallic conductors since an externally applied electric field can readily accelerate electrons into the many available empty states just above the Fermi level. In the case of the alkaline earths, the s-shell is filled. Metallic conductivity in these crystals arises from the fact that the s and p bands are sufficiently broadened at the equilibrium interatomic spacing that a small degree of overlap exists, so that electrons in Ca, for example, may be continuously excited from the $4s$ band to empty states in the $4p$ band.

Diamond and silicon provide examples of alternative behavior. In the case of diamond, the high degree of crystal symmetry and the large strength of the carbon-carbon interaction cause the $3p$ levels to split into not one, but two bands. There is about 5.5 eV separation between the top of the lower $2p$ band which corresponds to two states per atom and the bottom of the upper portion which derives from the remaining four (empty) $2p$ states per carbon atom. Diamond is, of course, an excellent insulator owing to the presence of this large forbidden gap separating filled from empty states. In the case of Si the interatomic interaction is much weaker and the forbidden gap is only 1.1 eV at room temperature. In pure Si a small but finite number of electrons are thermally excited across this gap from the valence band (the highest filled band) to the conduction band (the lowest empty band), and these are the source of electrical conductivity intermediate between that of metals ($\sim 10^6$ ohm^{-1}cm^{-1}) and insulators ($< 10^{-1}$ ohm^{-1}cm^{-1}).

As a practical matter, perfectly pure crystals are never obtained. The small amounts of impurity atoms present may affect the conductivity of semiconductor crystals significantly if they contribute electronic states which lie within the forbidden gap of the host crystal. In particular impurity atoms which differ in number of valence electrons from the host atoms will have either extra electrons, or too few, to satisfy the normal bonding arrangement and will thus tend to *donate* or *accept* these electrons to or from the crystal system, respectively. In the case of a *donor* impurity (one with an extra valence electron) there is a strong tendency for the extra electron to go into the conduction band, since much less energy is required to elevate these extra electrons from their bound states in the donor atom's Coulomb field to the conduction band than to elevate electrons from the valence band (bonding) states to the conduction band. At sufficiently low temperatures, of course, the Coulomb field will dominate and these extra donated carriers will freeze out on the donor atoms. The case of *acceptor* impurities, with one less electron than needed for bonding, is analogous. These tend to remove electrons from the host crystal, creating empty states at the top of the valence band. These empty states provide charge carrier propagation under the influence of an electric field much as do the carriers at the bottom of the conduction band.

2.2 Band Theory

The qualitative ideas of the preceding section may be placed on a quantitative basis. The quantum mechanical treatment of the electron states in a solid predicts the band structure described above, and provides much more insight into the nature of the electronic states. It is also necessary for a detailed understanding of the interaction of light with semiconductors. It is, however, mathematically complex, particularly if carried to calculation of actual band structures of real crystals. We will summarize some salient results in this section and indicate in a qualitative way the important concepts.

Two features of real crystalline solids which permit a solution for the allowed electronic states are the fact that a very large number of atoms are involved ($\sim 10^{22}$) and the fact that these atoms are arrayed, by and large, in a perfectly regular and spatially periodic way. The approximation may then be made that the number of atoms is infinite and that every atomic nucleus is located at one of the periodically spaced lattice positions. The potential energy function describing the nuclear attraction for the electrons is thus also a periodic function of the spatial coordinates having the same symmetry as the crystal itself. The solution wave functions describing the electron states must also have this property. This requires that the allowed electron states be representable by wave functions of the form

$$\psi(\vec{r}) = \phi_{\vec{k}}(\vec{r})\exp(i\vec{k}\cdot\vec{r}) \tag{2.3}$$

where $\phi_{\vec{k}}(\vec{r})$ has the same spatial periodicity as a function of position $\vec{r}$ as the crystal lattice. The set of vectors $\vec{k}$ define a reciprocal lattice array; they satisfy the relation $\vec{k}\cdot\vec{l} = 2\pi n$ where the displacement vectors $\vec{l}$ represent those rigid translations of the crystal as a whole which result in no change because of the crystal periodicity. Thus $\psi(\vec{r}+\vec{l}) = \psi(\vec{r})$. For example, for a one-dimensional potential array of wells spaced by a as shown in Fig. 2.2, the values of $\vec{l}$ would be just l

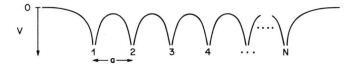

Figure 2.2 One dimensional array of N potential wells with constant spacing a.

$= na$ with n taking on integral values from one to the number (N) of atoms in the array, and k ranges from $2\pi/a$ to $2\pi/Na$ in value. Because of this periodicity, values of k and $k + 2n\pi$ are equivalent. In the limit $N \to \infty$, the range of k may be considered continuous from $k = 0$ to $1/a$ and may be identified with the momentum of the electron in the state $\psi_k(\vec{r})$.

The array of one-dimensional square wells represents a case for which the Schroedinger equation may be solved exactly. The condition that the energy eigenvalues be real restricts the allowed states to certain *bands* which may be compared to the discrete levels of the constituent square wells as in Fig. 2.3. Between these bands are forbidden gaps within which the energy eigenvalues are imaginary. Within the allowed bands the energy is a function of k, as shown in Fig. 2.4. Near the band edges the energy varies like

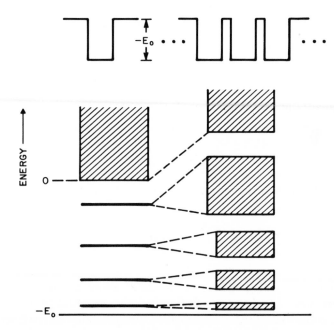

Figure 2.3 Energy bands of one dimensional array of square wells and development from discrete levels of isolated square potential well.

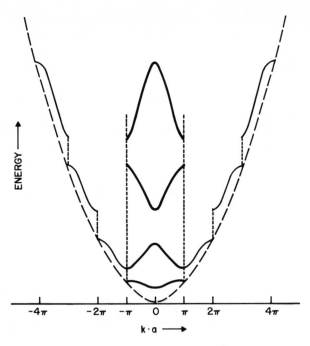

Figure 2.4   Energy as a function of momentum vector $\vec{k}$ within one dimensional array of square well potential periodically spaced by a, as in Figure 2.3.

$$E_k \sim E_n + A_n\left(k - n\frac{\pi}{a}\right)^2 \tag{2.4}$$

Such a quadratic dependence on k is reminiscent of that for a free particle of mass m^*

$$E_h = \frac{\hbar^2 k^2}{2m^*} \tag{2.5}$$

With the effective mass m^* defined by

$$m^* = \frac{\hbar^2}{\left[\dfrac{d^2E}{dk^2}\right]} \tag{2.6}$$

we see from Fig. 2.4 that near the edges of allowed bands m^* is roughly a constant for each band, positive near the band bottom and *negative* near the band top. The significance of a negative effective mass may be seen from Newton's first law: the particle will move backwards under an externally applied force. For charged particles in electric fields, this is tantamount to changing the sign of the charge, so that the motion of an electron in a state near the top of an allowed band under external electric fields can be expected to resemble that of a *positively* charged particle.

The generalization of these ideas to three dimensional crystals is straightforward but excessively complex to warrant detailed discussion here. The solutions resemble the one-dimensional case in that they consist of a plane-wave-like part and a part resembling an atomic wave function. Allowed and forbidden bands result, and some of the bands derived from different atomic states may overlap. Near the bottom and near the top of allowed bands an effective mass may be defined as in the one dimensional case and a quadratic dependence of energy on particle momentum is predicted in these regions.

The last filled band in a three dimensional crystal arises from the atomic states associated with the valence electrons of the atoms comprising the crystal. If this band is only partially filled (as with the alkali metals) or overlaps with an empty band (as with the alkali earths) a metallic conductor results. When the band is completely filled *no* conductivity results. (This may be considered the result of having, for every filled state of positive effective mass, a filled state of negative effective mass, so that the motion of the electrons in a filled band under an external field cancels pairwise and no conduction of current results). For real three dimensional crystals the structure of the electronic bands is a function of the *direction* as well as magnitude of $\vec{k}$, and the band minima and maxima need not occur at the center $(k_{x,y,z} = 2\pi \frac{n}{a}, \quad n = 0, \pm 1, \pm 2...)$ or edge $(k_{x,y,z} = \pi \frac{n}{a}, \quad n = \pm 1, \pm 3...)$ of the Brillouin zone. Portions of the band structure for GaAs, in which the top of the valence band and bottom of the conduction band occur at the center of the Brillouin zone, and for Si, in which the extrema occur at different values of k, are shown in Figs. 2.5 and 2.6. This is an important difference when the interaction of light with a semiconductor is to be considered, as we do now.

Energy and momentum must *both* be conserved for light to be absorbed by a semiconductor crystal. Photons in the near visible range have energies of 1 to 3 eV and momenta $\hbar k$ corresponding to $k = 2\frac{\pi}{\lambda} \approx 2\pi \times 10^4 \, \mathrm{cm}^{-1}$. For comparison the range of electron momenta is from $k = 0$ (zone center) to typically $k \approx 2\pi \times 10^8 \, \mathrm{cm}^{-1}$ at the zone edge, so that the momentum associated with visible photons is practically negligible on the scale of electron momenta. Thus direct electronic transitions involving only the absorption of a visible

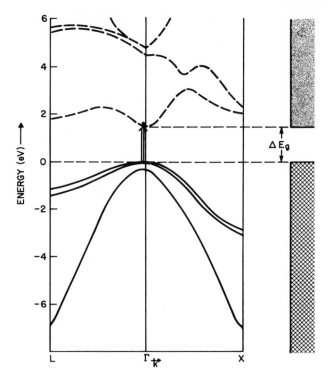

Figure 2.5 Band structure of GaAs (cubic symmetry). Filled levels (cross-hatched) and empty levels are separated by a gap, ΔE_g. The Γ point (cube center) corresponds to $\vec{k} = 0$. X and L correspond to $\vec{k}$ at its maximum value toward the cube face and toward the cube edge, respectively.

photon will occur nearly vertically on a diagram such as that in Fig. 2.6. For light with energy less than the band gap energy, a semiconductor will be transparent since there are no allowed electronic states which permit energy conservation.

For the case of Si, even for light with energy greater than the minimum energy spacing between conduction and valence bands, light absorption will still be weak because of the need to satisfy momentum conservation. This requires that an excitation which carries $\sim 10^8\,\mathrm{cm}^{-1}$ of momentum but very little energy be created or absorbed simultaneously with the photon absorption. The acoustic vibrational excitations of the lattice ions satisfy this criterion, as they have energies of a few milli-electron-volts and momenta up to that corresponding to the Brillouin zone edge. Thus absorption of light can take place but with reduced probability since it is now an *indirect* process. Materials such as Si are called *indirect band gap* materials for this reason. For sufficiently

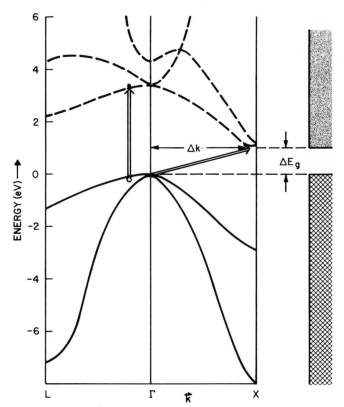

Figure 2.6   Analog of Fig. 2.5 for Si. The minimum energy transition across ΔE_g now corresponds to a large change in $\vec{k}$: an *indirect* bandgap.

increased photon energies *direct* transitions also become possible for Si as shown in Fig. 2.6. The strength of the optical absorption in a semiconductor thus depends strongly on whether the wavelength of the incident light is such that direct, or indirect, electronic transitions are possible. For direct transitions the extinction coefficient will be of the order of unity so that the light is mostly absorbed within a few optical wavelengths within the semiconductor, while for indirect transitions the light will penetrate 10 to 100 times as far. The absorption spectra of GaAs and Si illustrate this and are shown in Fig. 2.7.

Following absorption of a photon, an electron is present in the previously empty conduction band and a hole is present in the previously filled valence band. The filled valence band does not contribute to charge conduction since each occupied electron state is paired with another of equal and opposite momentum. The hole may be represented as a superposition of electron states near the top of the

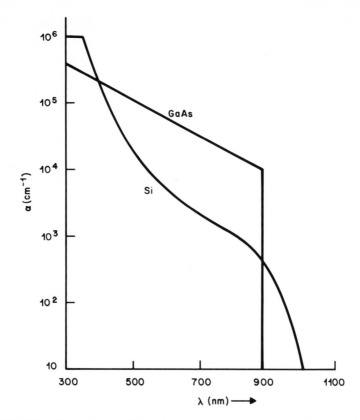

Figure 2.7 Absorption coefficient as a function of wavelength in Si and GaAs.

band, and since one is now empty, net nonzero momentum can result. an energy-momentum relation characterized by a negative effective mass. This is usually thought of as a positively charged carrier (hole) with positive effective mass.

Thus the electrons and hole each may function as a charge carrier and will respond to applied fields so as to give a current flow, with the hole and electron moving in opposite directions. If a built-in field is provided within the semiconductor, an electric photo-current will be available to flow through an external circuit. In the next sections we discuss the various ways this can be realized.

2.3 *pn* Junctions [2]

For the remainder of this chapter it will suffice to treat the electrons and holes in a semiconductor as a free electron (or hole) gas characterized by a constant effective mass m^*_e (or m^*_h), and a density or carrier concentration n (or p), for electrons in the conduction band (or holes in the valence band). These electrons or holes exist partly as a result of thermal excitation of electrons across the band-gap (intrinsic carriers), and partly as a result of thermal ionization of donor and acceptor impurities (extrinsic carriers). For reasons to be discussed in the last section of this chapter, semiconductors of interest for solar cell applications have band gaps of the order of one eV or more, which implies intrinsic carrier concentrations typically less than 10^{11} cm^{-3} at room temperature. Useful conductivity requires carrier concentrations orders of magnitude larger than this, so that the vast majority of the carriers are derived from ionized donor or acceptor impurities. These impurities are chosen to yield states which bind carriers weakly (~ 50 mev or less) so that at room temperature essentially all donors or acceptors are ionized and the number of free electrons or holes is very nearly equal to the number of donor or acceptor impurity atoms.

The ionized impurity atom bears, of course, an equal and opposite charge to that of the freed carrier, so that the semiconductor remains electrically neutral as a whole. The charged ions represent local perturbations of the periodic lattice potential and can scatter the free charge carriers. The carriers also scatter from the lattice of host atoms, which at finite temperature are no longer localized precisely at their ideal positions but undergo thermally excited vibration about those positions. The host lattice may contain defects of a nonimpurity nature, such as dislocations or other growth defects which break the translational symmetry of the perfect crystal. Vacancies, although strictly speaking a form of nonimpurity defect, act electrically just as a donor or acceptor impurity, and in some of the compound semiconductors may be the dominant "impurity" species observed.

Scattering of the charge carriers by the lattice vibrations (phonons) or ionized impurities limits the conductivity that can be obtained for a given carrier concentration. When an electric field $\vec{E}$ is applied the carriers are accelerated for the time between scattering collisions τ, attaining on average a velocity along the field

$$\bar{v} = -e\vec{E}<\tau> = -\mu\vec{E} \qquad (2.7)$$

where μ is defined to be the mobility of the charge carriers in that sample in terms of the average scattering time $<\tau>$. The probability for scattering by ionized impurities or lattice phonons may be separately calculated. It is usually assumed that the two types of scattering are independent so that the total scattering probability is the sum of the two; this gives

$$\frac{1}{\mu_{TOT}} = \frac{1}{\mu_I} + \frac{1}{\mu_L} \qquad (2.8)$$

for the resultant mobility μ_{TOT} in terms of μ_I and μ_L, the impurity and
lattice limiting mobilities. Actual mobilities can be determined from
measurements based on the Hall effect, and are less than μ_{TOT} in real
crystals because of growth defects. The typical variation of μ_I, μ_L and
μ_{TOT} with carrier concentration is shown in Fig 2.8 for GaAs and AlAs
at room temperature [3].

The net carrier concentration depends on the net *difference* between
the number of ionized donors and acceptors, N_A-N_D, while the impurity scattering limitation on μ_I varies as the *sum*, N_A+N_D. Hence for
high mobility it is important that *only* donors or *only* acceptors be introduced, and not both, so that the degree of *compensation* is small. Some
compound semiconductors can only be doped *p*-type, and others only
n-type, because of self-compensating effects. For example, the introduction of potential acceptor impurities (e.g., Li) into CdS does not
produce a *p*-type semiconductor because vacancies which act as donors

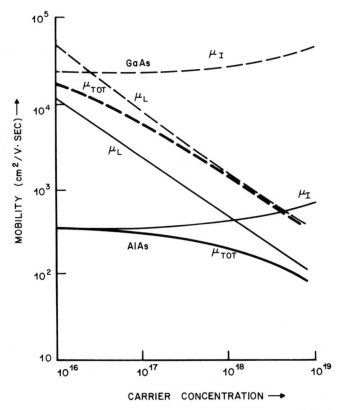

Figure 2.8   Room temperature mobilities for electrons in GaAs (dashed) and AlAs
(solid); μ_I, μ_L and μ_{TOT} as defined in text. From Ref. 3.

form in direct proportion to the acceptor impurity concentration and only a compensated, *n*-type material with reduced carrier mobility and hence higher resistivity results. Table 2.1 lists a number of semiconductors, typical donor or acceptor dopants used, and the range of resistivities attainable at room temperature.

A *pn junction* results when *p*-type and *n*-type semiconductors are joined. Conceptually we may think of bringing two slabs of opposite conductivity type into intimate contact. In practice such a junction is formed by growing a sulfur doped layer of, for example, GaAs on Zn-doped GaAs substrate; or by diffusing boron into phosphorus-doped silicon, giving the structure shown schematically in Fig. 2.9. In the former example an *abrupt np* junction is formed since there is a step-wise transition from *n*-type epilayer to *p*-type substrate, while the diffused junction gives a more gradual transition which can often be approximated as *linearly graded* for short distances around the *pn* transition.

Considering the situation in Fig. 2.9, it is apparent that some of the excess of electrons present in the *n*-region beyond those needed for bonding will tend to flow into the *p*-type region into bonding orbitals. This will result in the *n*-region acquiring a net positive charge, and the *p*-region a net negative charge, establishing a potential difference

Table 2.1

Typical Donor and Acceptor Dopants and Representative (low) Resistivities Obtainable at 300K for Various Semiconductors

Material	Donor	ρ_n(ohm·cm)	Acceptor	ρ_p(ohm·cm)
Si	As,P	10^{-2}	B,Al,Ga	10^{-2}
Ge	As	10^{-2}	Al,Ga	10^{-2}
GaAs	Si,Sn,S, Se,Te	10^{-3}	Ge,Zn, Cd,Be	10^{-2}
GaP	Si,Sn,S, Se,Te	10^{-3}	Zn,Cd, Be,Mg	10^{-2}
AlAs	[A/B]^a,Sn,Se	10^{-2}	Zn	.05
InP	Si,Sn	10^{-3}	Zn,Cd	10^{-2}
CdS	[A/B],I	10^{-3}		
CdSe	[A/B],I	10^{-3}		
CdTe	[A/B],I		[A/B],Cu,Li	10^{-2}
ZnTe			[A/B],Cu,Li	10^{-2}

^{a}[A/B] symbolizes controlled variation of stoichiometry (ratio of element A to element B) during or after growth.

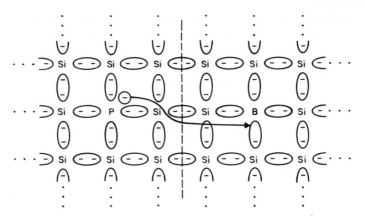

Figure 2.9   Schematic representation of atomic structure of a junction between n-Si (P-doped) and p-Si (B-diffused). Excess electrons on the P-doped side will tend to transfer to lower energy states in the incompletely filled bonding orbitals on the B-diffused side of the junction as shown by the arrow.

between the two. The charge transfer will stop, on balance, when this potential difference is equal to the initial difference in Fermi levels between the n and p regions. In the immediate vicinity of the junction the two semiconductor materials are depleted of free carriers, and an electric field (potential gradient) exists in the *space-charge* or *depletion layer,* as shown in Fig. 2.10.

In the one-dimensional approximation that the junction is planar and infinite in extent the junction space-charge ρ_{sc}, electric field $\vec{E}$, and potential distribution V may be simply calculated from Poisson's equation

$$\nabla^2 V(z) = -4\pi\rho_{sc}(z) \qquad (2.9)$$

to yield the results in Fig. 2.11. If contacts are made to the bulk of the semiconductor regions away from the junction, no external potential will be measured. This situation, with the Fermi level having a constant relative value across the junction, corresponds to a short-circuit external boundary condition. When an external voltage is imposed across the junction region, asymmetric conduction (a rectifying current-voltage characteristic) will be observed.

Under *forward bias* (p-region positive) the external potential serves to reduce the barrier to further flow of electrons into the p-region or holes into the n-region, and as the external voltage approaches the band-gap value, a rapidly increasing positive current will flow. In the

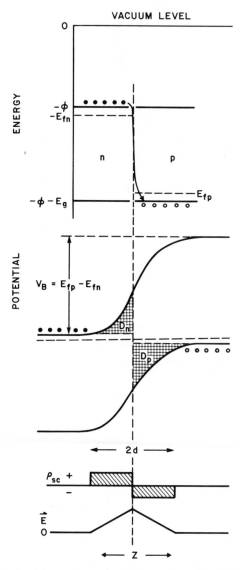

Figure 2.10 Energy-band representation of the *pn* junction and charge transfer kinetics shown in Fig. 2.9. Top: as *p* and *n* regions are joined, electrons flow to equalize free energy (given by Fermi level, E_F), This occurs (middle) when the electrostatic potential arising from transferred charge (lower) just offsets the initial Fermi level difference. Junction electric field (bottom), for a symmetric *pn* junction (bulk carrier concentration on *n* and *p* sides equal).

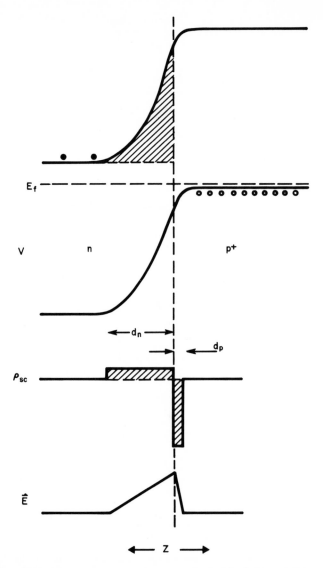

Figure 2.11 Potential energy, space charge, and electric field variation with position near an np^+ junction (p carrier concentration $>$ than on n side).

opposite case of reverse bias (p-region negative) the potential barrier at the junction is *increased* and only a small negative current will be observed. The functional form of the current is exponential:

$$I = I_o \left[\exp\left(\frac{eV}{nkT}\right) - 1 \right] \qquad (2.10)$$

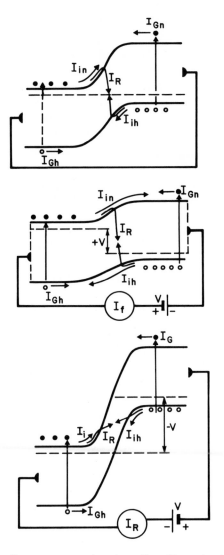

Figure 2.12 Current flow across a *pn* junction. Top: Short circuit, in the dark; generation currents $I_{Gp,n}$ of holes and electrons and injection currents $I_{ip,n}$; recombination current I_R. $\sum_{n,p} I_G + I_i = 0$. Middle: under external forward bias and Bottom: under external reverse bias.

At large reverse voltage I tends toward $-I_o$, the *reverse saturation current*. Even though solar cells operate under forward bias, this parameter is of importance since, as we shall show later in this chapter, the value of I_o determines the open-circuit voltage available from an illuminated solar cell. The parameter n is called the *diode factor* and determines (together with the absolute temperature T) the shape of the $I-V$ characteristic. Theoretical values of I_o and n may be calculated for various models. At finite temperature one must always have a diffusion component of the current; this ideal case gives the lowest possible value for n of unity. Electrons and holes may also recombine within the space charge region of the junction; this effectively reduces the barrier height by half and gives $n = 2$. These current processes are shown in Fig. 2.12.

Real *pn* junctions have current-voltage characteristics which are often well described by Equation 2.10 with values of n between 1 and 3, but with values for I_o somewhat larger than would be expected either for pure space-charge recombination or thermal diffusion limited current flow. Nearly ideal diffusion limited current-voltage behavior *can* be obtained in lightly doped small area silicon diodes, but this is not generally observed for the larger areas and heavier doping levels dictated for practical silicon solar cells. In compound semiconductors *pn* junctions tend to be even less well behaved. Tunnelling conduction involving defect states is often inferred, although the precise processes contributing to current flow are not generally clear in detail.

2.4 Metal-Semiconductor Junctions [4]

Two types of metal-semiconductor junctions are of importance for solar cells. Connections must be made between the semiconductor material and external metallic leads for current collection, and it is important that these junctions offer as little impediment to current flow as possible. In general a metal-semiconductor interface will exhibit rectification to a greater or lesser degree and special treatment must be used to assure a low resistance contact. The rectifying property results from a similar type of band-bending to that which gives rise to the junction field in a *pn* junction as discussed in the preceding section. This field may also be used to separate photo-generated electrons and holes. A metal-semiconductor junction which has rectifying characteristics is called a *Schottky barrier,* and for a given semiconductor, the barrier height may be maximized by proper choice of metal and processing. Very thin metal films are largely transparent to visible light yet still retain sufficient conductivity to make Schottky barrier solar cells practical.

In Fig. 2.13 the various possible potential energy diagrams for a metal-semiconductor junction are sketched schematically. In principal a metal with sufficiently low work function should make low resistance

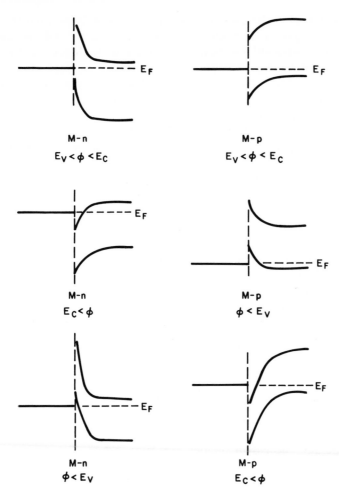

Figure 2.13 Possible energy band diagrams at various combinations of metal/semiconductor junctions. Top: surface depletion layer, rectifying (for n and p semiconductors). Middle: surface accumulation layer (ohmic contact). Bottom: surface inversion layer, pn junction induced within semiconductor. E_c, E_v and ϕ are the energies of the conduction band, valence band, and metal work function relative to the vacuum level.

nonrectifying (ohmic) contact to an n-type semiconductor, while a high work function metal should make low resistance ohmic contact to a p-type semiconductor. Maximum rectification would be expected for the reverse of these combinations. These ideal properties are not observed, as a rule, because the presence of the semiconductor surface has tacitly been ignored in Fig. 2.13. The atoms in the first few layers of the crystal do not occupy the same lattice positions as those in the bulk. A partial monolayer of atmospheric contamination, typically partial oxidation, is also expected for real semiconductor surfaces. Both of these effects can be shown theoretically to introduce surface electron states *within* the forbidden gap. These allowed states at the surface cause the Fermi level at the surface to be located near the middle of the energy gap. Metal-semiconductor junctions thus generally exhibit barrier heights between one-third and two-thirds of the bulk semiconductor energy gap, rather than the zero to full gap values suggested in Fig. 2.13.

When a *pn* junction solar cell is illuminated, electrons must be collected from the *n*-type region and holes from the *p*-type regions. The rectifying properties of metal contacts to these regions are such that this is the easy direction of current flow, the condition corresponding to *forward bias*, or injection of carriers over the barrier. However, if this barrier is a substantial fraction of the band-gap energy the maximum power obtainable from the cell will be significantly reduced. Efficient conduction *through* the barrier may be obtained by tunneling if the barrier width is reduced. This can be effected by doping the semiconductor surface region heavily so as to achieve a shallow depletion layer. Alternatively or additionally, excess surface states may be introduced simply by increasing the surface area by mechanical abrasion. Some of these excess states will lie within the barrier and render it leaky. Both heavy surface diffusion of donor or acceptor impurities and mechanical abrasion can result in a deterioration of the underlying junction if not carefully performed. In practice the attainment of low resistance metal-semiconductor contacts is a matter of art rather than science, and even generally successful recipes are often prone to poor reproducibility.

The maximum Schottky barrier height for a given metal semiconductor junction is limited to a value less than or equal to that for a *pn* junction in that semiconductor. To obtain or approximate this value the semiconductor surface must be treated to give a structure for which the Fermi level of the surface is at the same energy value as in the bulk; i.e., the density of surface states lying within the band gap must be reduced to a negligible value. In order to achieve this, the nature of the semiconductor surface may be modified by vapor or liquid phase chemical treatment under various oxidizing or reducing conditions. In general a metal-semiconductor junction is actually a metal-insulator-semiconductor sandwich. The purpose of the various empirically determined premetallization surface treatments is to convert the interfacial insulator layer to the appropriate conductive or semiconductive form

Table 2.2

Barrier Heights on n-Type Semiconductors and Band Gap Energy (eV)

	Si	Ge	GaAs	CdSe	InP
Al	00.76	0.48	0.8		
Ag	0.79		0.88	0.43	0.54
Cu	0.79	0.48	0.85	0.33	0.48
Pt	0.90		0.85	0.37	0.51
Au	0.83	0.45	0.88	0.49	0.49
E_g	1.11	0.66	1.43	1.67	1.35

which will yield either a maximum or minimum barrier height. The barrier heights for several metal-semiconductor combinations are listed in Table 2.2; it is apparent that *pn* junctions give substantially higher barrier voltages in all instances. In spite of this inherent limitation, the intrinsic simplicity and potential ease and cheapness of manufacture of Schottky barrier cells has kept interest in this approach alive, at least at the research level (see Chapter 4).

2.5 Heterojunctions

A *heterojunction* is formed between two different semiconductor materials. *Pp*, *Nn*, *Pn* and *pN* combinations (the upper case letter conventionally denotes the semiconductor partner with larger band gap) all may show rectifying current-voltage characteristics, and at least in principle all could provide usefully large photoelectric response. Heterojunctions offer several advantages over homojunction or Schottky barrier cells. While the full barrier height (of the lower band gap partner) is retained, the wide-gap material may be chosen to be largely transparent to the solar radiation so that the light is all absorbed in the low-gap material. The photo-generated minority carriers thus do not contact the surface and hence cannot recombine through surface states. There may, of course be *interface* states associated with the heterojunction, which, like surface states, may lie within the band-gap. These would not only permit recombination of photo-generated carriers but would cause deterioration of the heterojunction current-voltage characteristic. Accordingly, it is essential that as low a density of interface states as possible be attained if the full potential advantage of heterojunction solar cells is to be realized.

A minimum interface state density is realized when the semiconductor pair comprising the heterojunction has the same crystal structure and relative arrangement of atoms. This is most simply realized when both materials have the same crystal symmetry and closely equal *lattice constant* or unit cell size, as in the case of the AlAs/GaAs pair (both have the zincblende cubic structure; the room-temperature lattice constants are 5.661 and 5.654 Å respectively). A good match may also be realized for certain orientations of other pairs, e.g., the atoms on the (0001) face of CdS (wurtzite, hexagonal structure) are spaced to match closely the atoms of the (111) face (perpendicular to the cube diagonal) of InP (zincblende structure). Another alternative is the *heteroface* structure, in which the *pn* junction is located *within* the low band-gap partner, at a small distance from the heterointerface, giving a *Ppn* or *Nnp* structure. This compromise removes the influence of heterointerface states on the junction current-voltage characteristic, but minority carrier recombination through interface states remains as a potential problem.

An additional consideration of possible importance involves the structure of the bands within the depletion region of an abrupt heterojunction. A spike-notch structure will result for either the *Pn* or the *Np* combination, or both, as shown in Fig. 2.14. Interdiffusion during growth of the heterojunction may wash out this structure, but comparison of theory with experiment on this point is rendered difficult by the necessity to measure the relative energies of the conduction and

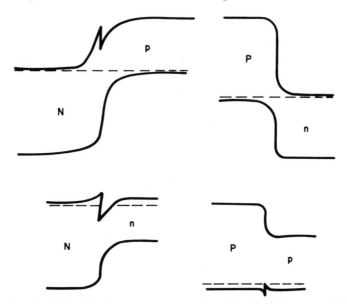

Figure 2.14   Schematic representation of spike-notch structure at abrupt heterojunction as exemplified for AlAs-GaAs combination. (From Ref. 5 and 6).

valence bands of the two materials. In comparison to the energy of the band gap, which can be accurately determined from optical absorption measurements, the *electron affinity,* or energy of the (bulk) conduction band minimum from the vacuum level, is not an experimentally well defined quantity. Photoelectric emission measurements of semiconductor work functions are highly sensitive to surface conditions, even for nominally clean surfaces, in contrast to such measurements on noble metal surfaces. The experimental evidence that "spike-notch" effects are important for any actual heterojunction is probably strongest for the AlAs/GaAs heterojunction pair, but alternative interpretations, even in that case, are possible [5,6].

A final heterojunction structure of interest should be mentioned. There exist several transparent, highly conducting oxides such as SnO_2, ITO (indium-tin-oxide, typically 90 percent SnO_2, 10 percent In_2O_3) and Cd_2SnO_4 which are in fact degenerate n-type semiconductors with $n \sim 10^{20}cm^{-3}$. A junction between one of these and a more lightly doped n-type semiconductor results in a Schottky barrier. Such a junction to a p-type semiconductor gives rise to either Schottky barrier or *pn* junction behavior, depending in part on whether the fabrication technique gives rise to interdiffusion of the two semiconductor species. The possibility that a metal-insulator-semiconductor (MIS) structure is formed seems particularly likely in this case, as a probable result of oxidation-reduction reaction at the conducting oxide-semiconductor interface.

A low-resistance ohmic contact between conducting oxides such as ITO and semiconductors is of interest for the formation of transparent conductive windows to improve current collection from solar cells without obstructing the area exposed to sunlight. In this case it is particularly important that chemical reaction between the window material and the semiconductor not result in formation of a thin insulating layer.

The theoretical understanding of heterojunction characteristics appears to be well-developed, but it is difficult to apply to particular material situations because of uncertainties regarding the details of the heterointerface under specific experimental conditions. In general the following considerations apply:

1. A close lattice match (less than 1 percent difference in lattice constants and the same crystal structure) is required for good *Pn* or *Np* heterojunction performance.

2. A close match in thermal expansion coefficient, at least over the range of fabrication and operation, is necessary to ensure the physical and mechanical integrity of the heterojunctions.

3. The pair should be chemically stable against formation of other phases. AlAs and SnO_2 react to form an insulating interfacial oxide (presumably due to formation of Al_2O_3), for instance.

4. If the growth process is expected to result in an atomically abrupt heterojunction, proper account of the electron affinities for the semiconductor pair must be taken so that undesirable spike or notch impediments to carrier transfer are avoided.

2.6 Idealized *pn* Junction Solar Cell

The underlying principles of current and voltage (and hence power) generation in solar cells may be understood from analysis of the effect of illumination on the simplified *pn* homojunction structure of Fig. 2.15. We assume light incident on the *p*-type face. Photons having energy greater than the band gap energy E_g will be absorbed; this absorption follows Beer's law which states that the light intensity a distance x into the material $I(x)$ has been reduced from the incident intensity I_o as

$$I(x) = I_o \exp(-\alpha(\lambda)x) \qquad (2.11)$$

where $\alpha(\lambda)$ is the (wavelength dependent) absorption coefficient. Typically α is of order $10^4 - 10^5 \, cm^{-1}$ for direct gap semiconductors and of order $10^3 \, cm^{-1}$ for indirect gap semiconductors as discussed earlier in Section 2.2. As the light is absorbed, electron-hole pairs are created in the three regions shown : the *p*-type front face, the depletion region

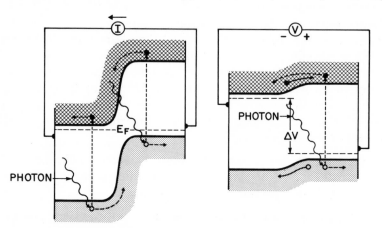

Figure 2.15   Band diagram and current processes for illuminated *pn* homojunction. Left: short circuit ($V=0$) and right: open circuit ($I=0$) boundary conditions.

around the junction, and the n-type base. An ohmic back contact is assumed. This is essentially metallic in optical properties, and will absorb long wavelength photons which have not created electron hole pairs in the semiconductor. These do not contribute to the overall current, however.

The electrons in the face and holes in the base diffuse toward the depletion region, and in that region are swept by the junction field across the junction and are collected as majority carriers on the other side. This results in an excess of electrons in the base, and of holes in the face. Under *open circuit* conditions, i.e., with no external electrical path *around* the junction, the face is charged positively and the base negatively. The effect on the junction is the same as an externally imposed *forward bias* and forward injection current will flow. A new equilibrium will be reached when the voltage V developed across the junction is such that the injection current just cancels the photocurrent I_{ph}:

$$I_{ph} = I = I_o(\exp(eV/nkT) - 1) \tag{2.12}$$

This last relation is readily inverted to yield an expression for the open-circuit voltage

$$V_{oc} = \frac{nkT}{e} \ln\left[\frac{I_{ph}}{I_o} + 1\right] \tag{2.13}$$

While Equation (2.13) suggests at first sight that V_{oc} increases with n, the diode ideality factor, this is not the case. I_o increases so rapidly with increasing n that a maximum V_{oc} for given I_{ph} occurs for $n = 1$.

If a conductor is connected round the cell a current will flow, with value $I_{sc} = I_{ph}$ for a short circuit (zero load resistance) and a value I given by solution of

$$I = I_{ph} - I_o(\exp(e(IR)/nkT) - 1) \tag{2.14}$$

for an external load resistance R. An optimum value of load resistance exists such that the output power is a maximum. This may be determined by multiplying Equation 2.14 by the load voltage $V = IR$, differentiating with respect to V and setting equal to zero to give Equation 2.15:

$$\frac{I_{ph}}{I_o} + 1 = \left[1 + \frac{eV_m}{nkT}\right]\exp\left[\frac{eV_m}{nkT}\right] \tag{2.15}$$

This may be solved numerically for V_m, whence substitution into 2.14 gives I_m, and the optimum load resistance R_m is given by their ratio.

R_m depends on I_{ph}, I_o, and n and thus is a function of the intensity of the light as well as of the junction properties.

The ratio of the maximum power $I_m \cdot V_m$ to the product of V_{oc} and I_{sc} is called the *fill factor*. It represents the ratio of the largest area rectangle that can be inscribed in the fourth quadrant of the $I-V$ curve obtained under illumination to the area of the rectangle bounded by $I = I_{sc}$, $V = V_{oc}$ and the I and V axes, as shown in Fig. 2.16. These three parameters, V_{oc}, I_{sc} and fill factor (FF) are the essential descriptors of solar cell performance. The *power efficiency* is given by

$$N = \frac{(FF) I_{sc} V_{oc}}{incident\ solar\ power} \tag{2.16}$$

There are two primary sources of fundamental inefficiency in the operation of a solar cell. The first has to do with the mismatch of the semiconductor energy gap with the solar spectrum. Low energy photons, with $h\nu < E_g$, are not absorbed in the semiconductor where electron-hole pairs are created, but simply heat up the back contact. Photons with energy greater than E_g create electron-hole pairs, but the excess energy $h\nu - E_g$ is similarly lost in heating up the semiconductor. The second basic inefficiency mechanism relates to recombination of electron-hole pairs before the minority carrier crosses the junction. At best this occurs at the rate given by band-to-band radiative recombination, and in practice occurs at a faster rate because of additional

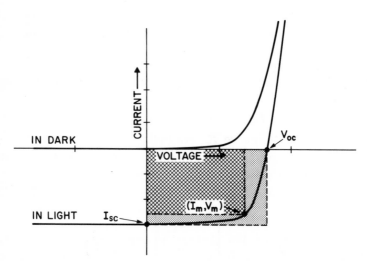

Figure 2.16 The fill factor is the ratio of the area of the cross-hatched rectangle to the larger shaded rectangle.

non-radiative recombination channels through bulk impurity centers, or at extended defects, interfaces, or free surfaces. This recombination not only represents a loss of carriers available to contribute to I_{ph} and hence a diminution of the current collection efficiency, but also determines the values of I_o and n and hence the voltage at which the current is available. Naturally there are also extrinsic sources of inefficiency as well. Some portion of the incident light is reflected from the semiconductor surface and is not absorbed for that reason. Finite resistance to majority carrier flow in either base or face regions contribute to nonzero series resistance. A contacting grid on the top surface reduces series resistance but causes partial shadowing of the cell. These factors are susceptible to improved materials technology and intelligent design and will be considered in later chapters in the context of particular cell structures. We will conclude this section with a consideration of the basic limitations on cell efficiency over which no engineering control is possible.

Just as electron-hole pairs are created by photon absorption, they may recombine with photon emission. These reradiated photons may be reabsorbed, or they may escape the solar cell and represent a loss to the current flowing to the load. The rate of radiative recombination is proportional to the product of electron and hole densities so that

$$J_{ph} = e\Phi - Anp \tag{2.17}$$

where Φ is the total flux of absorbed photons and A is a constant of proportionality which depends on the same properties of the semiconductor which determine the strength of optical absorption. Now the $n \cdot p$ product is bias dependent, specifically

$$np = n_i^2 \exp \frac{eV}{kT} = N_c N_V \exp \left| - \frac{E_g - eV}{kT} \right| \tag{2.18}$$

where N_c and N_V are densities of states in the conduction and valence bands. By combining the last two equations we get the diode equation (with $n = 1$) and I_o identified with

$$I_o = AN_c N_V \exp \left(- \frac{E_g}{kT} \right) \tag{2.19}$$

(This model of the junction ignores recombination in the depletion region (where $np \rightarrow 0$). If defects *at* the junction or imperfections in the depletion region are significant a similar calculation can be made which results in a version of the diode equation with $n = 2$.)

Equation 2.19 may be used to show that the voltage at maximum power output satisfies

$$E_g - eV_{max} = kT \ln\left\{\frac{An_c N_V}{I_{ph}}\left[1 + \frac{eV_{max}}{kT}\right]\right\} \qquad (2.20)$$

The current density collected is just

$$J_{ph} = \left\{1 + \frac{kT}{eV_{max}}\right\}^{-1} \Phi \qquad (2.21)$$

To proceed further we need to know A, N_c and N_V for the material of interest. As an example we can use values for GaAs at 300K [7,8]: $E_g =$ 1.423 eV, $A N_c N_V =$ 14,400 amp cm^{-2} and $e\Phi =$ 0.0296 amp/cm^2 (one sun, AM 1.0, flux absorbed for $h\nu > E_g$). This yields $E_g - eV_{max} =$ 0.433 eV, and $J_{ph}/e\phi =$ 0.975. Thus the effect of radiative recombination is a *small* 2.5 percent current loss and a *sizable* 28 percent voltage limitation at one sun intensity. The work per absorbed photon is then $eV_{max}(J_{ph}/e\phi) =$ 0.965 eV or 0.458 eV less than E_g. Thus for each photon absorbed (i.e., those with $h\nu > E_g$) only 63 percent of E_g in work is produced, and this under ideal conditions.

For other semiconductors, $E_g - eV_{max}$ is a slowly varying function of band gap and lies in the 0.4 to 0.5 eV range for $0.5 \, eV < E_g < 3 \, eV$. As E_g is increased, evidently fewer photons are absorbed but the available work per photon increases, so that an optimum does exist. For these assumptions the maximum is fairly flat and values of E_g between 1.2 and 1.5 eV provide a good match to the terrestrial solar spectrum. From Equation (2.17) it is apparent that eV_{max} approaches E_g as J_{ph} is increased so that the ultimate efficiency of a cell may be increased by focusing. Concentration of sunlight by several thousand times has been used to demonstrate this effect with GaAs heterojunction cells. The principal limits to this approach are series resistance, heating effects, and of course the complexity and cost of the high magnification optics. The optimum value of E_g scales with kT, so higher values of E_g are appropriate for the elevated temperatures associated with concentration or combined photo-voltaic/thermal energy systems.

The analysis of idealized heterojunction or Schottky barrier solar cells may be made analogously. For the heterojunction cells the light absorbed in the wide-gap window may or may not be important, but otherwise the analysis is identical to that here. For metal-semiconductor junctions nearly ideal Schottky barrier behavior ($n \rightarrow 1$) is often realized. MIS structures have been found to give enhanced barrier heights in many cases, however. The barrier heights E_B observed are generally dependent on processing details but are of course limited to the maximum value set by the band gap of the semiconductor. The decrease in available work per photon from E_B is determined ideally by bulk radiative recombination and is hence the same as that considered above for *pn* junctions, so that $E_B - eV_{max}$ will be less than that for a *pn* junction by the same amount E_B is less than E_g. This additional loss amounts typically to several tenths of an eV.

2.7 Practical Analysis

In evaluating a real solar cell, the measurements of interest are the values of V_{oc}, J_{sc}, FF and of course power efficiency. These are customarily made in accordance with standardized procedures, e.g., the "Terrestrial Photovoltaic Measurement Procedures" [9] established by the National Aeronautics and Space Administration of the United States government. As a rule, the efficiency so determined is significantly less than the theoretical limit for the materials of which the cell is made, and evaluation consists of further measurements and analysis to determine which particular characteristics of the cell represent weak links in the overall efficiency chain. In addition to the four single point variables listed above, the complete current-voltage characteristic, both in the dark and under illumination, and a plot of spectral response (relative short circuit current versus wavelength) provide the necessary data to determine whether bulk quality, junction quality, or external quality (contacts, anti-reflection coating, etc.) need improvement. The usual model used for analysis is based on the equivalent circuit in Fig. 2.17.

Physically, the sources of shunt conductance G_s are regions of reduced junction barrier height. Direct electrical shorts may arise at

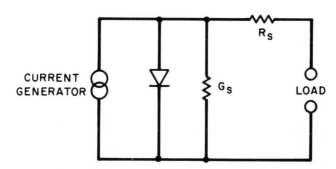

Figure 2.17 Equivalent circuit for a solar cell includes a current generator with output proportional to light intensity, a shunt conductance, series resistance, and voltage limiting diode.

pinholes or microcracks in the junction-forming layer or shorts in the contact metallization around the junction. The sources of series resistance include high contact resistance or nonohmic contacts, the bulk resistivity of the base layer, and spreading resistance in the thin front face layer. These imperfections affect cell performance most obviously through reduction of the fill factor, as shown in the linear current-voltage curves in Fig. 2.18. Specific values of G_s and R_s, together with the diode parameter J_o and n, may all be determined from log current-voltage plots as in Fig. 2.19. In extreme cases it will not be possible to obtain a well-defined value of n from the slope of the log $I-V$ plot. This may arise from excessive values of G_s and R_s or from an apparent voltage dependence of n. The latter can result from a high density of interface states at the junction giving rise to a dependence of barrier height on bias voltage, an effect rarely observed for pn junctions but rather typical of Schottky barrier and particularly MIS structures.

Measurements of relative spectral response can provide information about the quality of the bulk semiconductor region in which the light absorption takes place. The absorption coefficient for short-wavelength light is larger than for the longer wavelengths so the spatial distribution of minority carrier generation is a function of the wavelength of the incident light. If a semi-infinite geometry is assumed, one-dimensional diffusion equations may be written to describe the generation and flow of the minority carriers for a monochromatic excitation flux $\Phi(\lambda)$, using a diffusion coefficient D and effective lifetime τ;

$$D \frac{\partial^2 n}{\partial z^2} + \frac{n}{\tau} + \frac{\Phi(\lambda)}{\alpha(\lambda)} \exp(-\alpha(\lambda)z) = \frac{\partial n}{\partial t} \qquad (2.22)$$

D and τ may be combined to yield a diffusion length $l = D\tau$ and with appropriate boundary conditions to reflect open or short circuit conditions and cell geometry the short circuit current may be calculated.

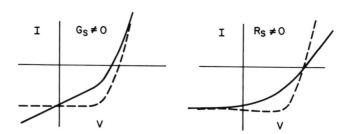

Figure 2.18 $I-V$ curves showing (left) the effect of shunt conductance and (right) the effect of series resistance, both of which reduce the fill factor in comparison to the ideal (dashed) case ($R_s, G_s \rightarrow 0$).

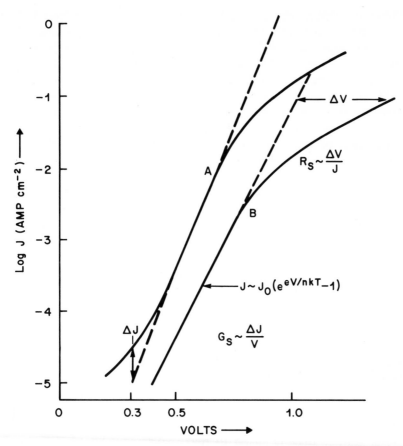

Figure 2.19   Log current density vs. voltage plots as measured in the dark for two GaAs diodes. J_o is determined from the value, and n from the slope, of the linear region; G_s and R_s from the offsets from linearity at low and high bias respectively.

For the simplest case where a negligibly thin (or optically transparent, as in a heterojunction) front face is assumed, the quantum efficiency may be shown to be just

$$\frac{J_{sc}(\lambda)}{e\Phi(\lambda)} = \frac{\alpha(\lambda)l}{1 + \alpha(\lambda)l} \tag{2.23}$$

Thus a value for effective diffusion length may be obtained from measurements of short circuit current density or relative quantum efficiency at several wavelengths without the need for the considerably more difficult and less accurate measurement of absolute quantum efficiency.

These measurements should be made at several intensities to verify that the diffusion length obtained is independent of illumination intensity. This may not be the case if long-lived trapping centers are present, for instance. Otherwise a broadband bias light source with the approximate spectral distribution and intensity of solar illumination should be used in conjunction with a differential measurement of spectral response such as can be obtained with a chopped lower-power monochromatic source and lock-in detection. Correction for the spectral variation of the cell antireflection coating performance should of course be included in relating the absorbed light flux $\Phi(\lambda)$ to incident intensity when applying equation (2.23).

For many cells with reasonable performance well-defined values of diffusion length, diode parameters, shunt conductance and series resistance can be obtained, permitting shortcomings in the performance parameters to be diagnosed. Thus contact problems can be separated from bulk-lifetime problems, for example, and effort can be directed toward improving material growth or processing technology as appropriate.

2.8 Summary

The fact that charge carriers in semiconductors exist in two bands separated by an energy gap permits their use in fabricating efficient solar cells. *Pn* homojunction, *Pn* or *Np* heterojunction, Schottky barrier, heteroface/homojunction, or MIS configurations may all be used. An important distinction is that between *direct* and *indirect* gap semiconductors. Either may be used to make solar cells, but a hundredfold greater thickness of *indirect gap* material is required to ensure maximum absorption of the incident solar radiation as compared to the few micrometers thickness of direct gap material needed.

The limiting efficiency for a semiconductor junction or metal-semiconductor junction solar cell is determined by the value of the energy gap. For terrestrial sunlight an optimum value of E_g exists, in the range of 1.2 to 1.5 eV, corresponding to a maximum power conversion efficiency of $\sim$28 percent at 300K. This is less than one-third the $\sim$95 percent efficiency of an ideal heat engine with 5800K input temperature and 300K exhaust.

The quantities of interest in specifying performance of a solar cell are V_{oc}, J_{sc}, fill factor and efficiency. Curves of the dark and illuminated current-voltage relation and short circuit current spectral response are important for analyzing cell imperfections. These permit separation of shunt and series resistance effects, often associated with external contact or junction fabrication problems, from intrinsic problems, such as reduced diffusion length and excessive density of bulk nonradiative recombination centers.

REFERENCES

1. The reader interested in a detailed treatment is referred to a standard solid-state physics test, such as C. Kittel, *Introduction to Solid State Physics, 5th Ed.* (Wiley, New York, NY, 1976), Chapters 1, 7 and 8.

2. For a detailed treatment see, for example, Sze, S.M. *Physics of Semiconductor Devices* (Wiley-Interscience, New York, NY, 1969), Chapter 3.

3. W. D. Johnston, Jr. and W. M. Callahan, *J. Electrochem Soc.* **123**, 1524 (1976).

4. The material in this and the succeeding section is covered in detail in A. G. Milnes and D. L. Feucht, *Heterojunctions and Metal-Semiconductor Junctions* (Acad. Press, New York, NY, 1972).

5. W. D. Johnston, Jr. *IEEE Trans.* **ED24,** 135 (1977).

6. H. C. Casey, Jr., A. Y. Cho, H. Kroemer and W. Y. Chien, *Appl. Phys. Lett.* **33,** 749 (1978).

7. D. D. Sell and H. C. Casey, Jr. *J. Appl. Phys.* **45,** 800 (1974).

8. H. C. Casey, Jr. and F. Stern, *J. Appl. Phys.* **47,** 631 (1976).

9. *Report #ERDA/NASA/1022-77/16 (NASA TM 73702)* (United States Nat. Tech. Information Service, Springfield, VA., 1977).

CHAPTER 3
Established Technologies

There are four photovoltaic materials approaches that can be considered established in the sense that the bulk of current effort devoted to them has moved beyond the research stage to development or production engineering. In order of the extent of commercialization or practical development these are single or semicrystal Si cells packaged for flat-plate arrays, single crystal Si cells for concentrator systems, CdS/Cu_2S thin film arrays, and single crystal GaAs cells for high-performance concentrator use.

In 1978 flat plate Si modules were available commercially from 10 manufacturers, with array efficiencies up to 13 percent at prices ranging from \$12.00 to \$25.00 per peak watt of rated power when purchased in quantity. By the end of 1979 this price had dropped to the \$8.00 to \$15.00 range. Single crystal Si cells for concentrator application and complete concentrator electric/thermal systems were also available on a custom order basis from several of the larger companies producing flat-plate arrays. Two companies had pilot-line production facilities for Cd/Cu_2S arrays in operation. Production arrays of this material are available but peak efficiencies have historically been less than 6 percent. GaAs concentrator cells and systems are not available commercially at this time, but concentrator modules with efficiencies well over 20 percent at concentration over 1000 times have been demonstrated. Arrays of such modules with output power of one to several kW peak are under construction.

In this chapter the fabrication technology, performance limitations and remaining hurdles for each of these more established approaches will be examined in turn.

3.1 Single or Semicrystal Si Cells

3.1.1 Si *Ingot and Wafer Production*

The manufacture of solar cells from sliced, polished wafers of single crystalline Si is the traditional approach to obtain usefully high efficiency, and results in the type of device shown in Fig. 3.1. The starting material is quartzite gravel or crushed quartz. It has been popularly stated that Si solar cells may be made from sand, but this is not usually a practical approach owing to the relative impurity of beach or desert sand in comparison to quartz rock, which is typically 99 percent SiO_2. This material is reduced with coke to yield metallurgical grade Si of moderate (98 percent) purity. The metallurgical grade Si is converted to $SiHCl_3$ and further purified by fractional distillation. The $SiHCl_3$ is then reduced with H_2 to yield semiconductor grade polycrystalline Si.

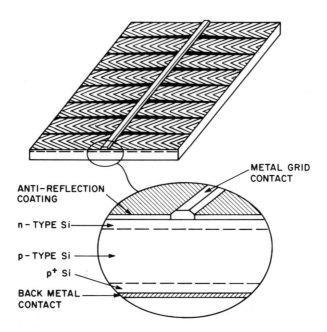

Figure 3.1   Construction of typical single crystal silicon solar cell.

This poly-Si is melted and converted to a single crystal boule or ingot. Traditionally this has involved either Czochralski or floating-zone technique. In the former, a seed crystal is lowered into the melt and then slowly withdrawn while being rotated (see Fig. 3.2a). Most of the common residual impurities in semiconductor grade Si are more soluble in the molten phase and tend to concentrate there as Czochral-ski growth proceeds. Large boules may be produced; they are usually rendered cylindrical by centerless grinding for subsequent ease of han-dling. Ten cm diameter and tens of cms length are commonplace in industrial production.

In floating-zone growth a localized molten region is produced and

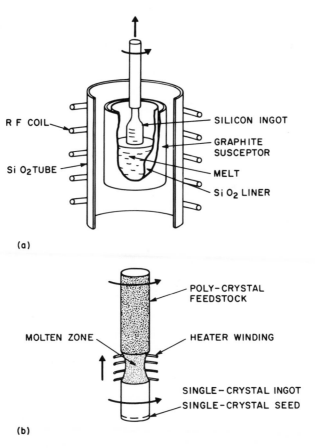

(a)

(b)

Figure 3.2 (a) Schematic drawing of Czochralski growth apparatus for production of single crystal Si ingot. (b) Schematic drawing of floating zone process for conversion of polycrystalline to single crystalline Si.

propagated along a rod of polycrystalline material, either by movement of the rod or the heat source, (see Fig. 3.2b). This is repeated several times in the same direction so that impurities are swept to one end. At the same time the crystallite size grows significantly and single crystal growth is readily obtained. Alternatively, a single crystal seed may be used, much as in Czochralski growth. The floating-zone process uses no crucible to contain the molten Si, which is retained by surface tension between the melting and growing solid Si faces. In Czochralski growth an SiO_2 crucible is generally used and oxygen contamination of the growing Si boule results. This is avoided with the floating zone technique. In principle there is no limit to the length of either Czochralski or floating-zone Si boules. The diameter of Czochralski boules and the maximum growth rate attainable are related by the necessity to transfer latent heat from the molten interface. Slow growth rates aid in rejection of impurities to the molten phase, however, and the maximum growth rate is not usually employed. These principles apply to floating-zone growth as well with additional constraints on boule diameter set by heat flow geometry (which is radial into the molten zone in this case) and the difficulty of maintaining a large liquid mass in the self-supported molten zone.

Following Czochralski or floating-zone growth the single-crystal boules or ingots are sliced into wafers, polished, and etched to eliminate work damage left from the sawing operation. A final thickness of 0.5 mm is typical for a 7.5 cm diameter wafer; proportionately thinner wafers of smaller diameter also have adequate mechanical strength to be handled without fracture. More than half of the single crystal Si is lost in this boule-to-wafer conversion. Moreover, the resulting wafers are circular and additional loss is involved if they must be sliced to the rectangular or hexagonal shapes required for the close packing of finished cells which yield efficient area utilization.

Actual fabrication of solar cells from the polished, sliced Si wafers proceeds through junction formation, contacting, and application of an anti-reflection coating. The cells are then bonded together and encapsulated or sealed into an environmentally protected module. A great part of the total energy or dollar investment goes into the traditional wafer fabrication and techniques to reduce these costs are under intense development. These include efforts to find alternatives to the $SiHCl_3$ refinement procedure as well as efforts to cast the molten Si directly into rectangular cross-section ingots or sheet, bypassing the consumption of time and energy expended in the Czochralski growth steps.

The representative cost of metallurgical grade Si is presently one dollar per kgm and there does not appear to be any problem of quartzite supply [1]. The annual U.S. production capacity is currently 200,000 metric tons (500,000 metric tons world wide) [2]. Polycrystalline semiconductor grade Si costs about $75.00 per kgm [3]. The production rate is limited by the trichlorosilane process which is inherently batch-oriented and slow. To place these figures in perspective, state-of-the-art multiblade or multiwire saws can produce one square meter of wafers per kgm of Si ingot [4]. Suitable ingot starting material must become available at $7.00 to $8.00 per kilogram if a wafer-price goal of $10.00/m^2 (~$1/ft^2) is to be met. This would allow another $10.00/m^2 for junction and contact fabrication and a final $10.00 for encapsulation to yield a price of $30.00/m^2 or $0.30 per peak watt at 10 percent array efficiency. Present semiconductor grade wafer cost [5] is ~$350.00 per m^2 ($3.50/peak watt) representing somewhat less than one-third of the present cost of encapsulated flat-plate modules.

It is not actually necessary that the Si wafers be strictly single crystal. Provided the grain size is large enough (several mm in practice) cells of good efficiency (>10% A.M 1.5) can be prepared. Such large grain poly-crystal material, or semicrystalline [6] material as it is called to distinguish it from low efficiency small grain polycrystal material, can be cast as rectangular ingots. The cooling rate must be fairly slow to ensure large grains but advantages over Czochralski growth are evident, both in ingot shape and in time and energy saved. The structure of this large grain, cast-ingot material consists of columnar, oriented grains [7]. A largely single-crystal ingot may be obtained by directional solidification from a single crystal seed using a heat-exchanger technique [8] originally developed to produce large (30 cm diam) boules of single crystal sapphire.

An outline of the heat-exchange directional solidification apparatus is shown in Fig. 3.3. A charge of polycrystalline Si is melted above a seed crystal which is cooled by a flow of the gas passing through the heat exchanger pedestal. The molten charge is contained in a graded SiO$_2$ crucible which has a fully dense interior surface and a low density exterior. During cooling the exterior delaminates and the thin inner layer breaks away so that the differential thermal expansion between crucible and ingot does not crack the ingot. The fact that the ingot grows upwards ameliorates the effects of convection at the solid-liquid interface and eliminates oxide impurities which float to the melt surface, such as SiO, which tend to segregate at the growth interface in Czochralski growth. No mechanical motion or rotation of the crucible, seed or heat-exchanger furnace components are required. This permits much simpler furnace construction and a potentially low maintenance cost. The SiO$_2$ crucible may be used only once for either the heat

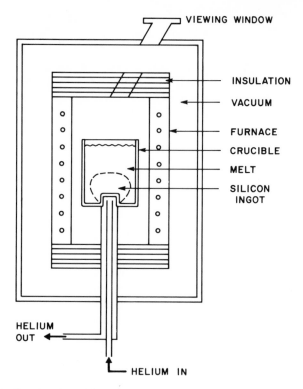

Figure 3.3   Outline drawing of heat exchanger apparatus for directional solidification of Si melt. (After Ref. 8).

exchange or Czochralski method and the cost of the crucibles may turn out to be the ultimate factor limiting cost reduction possibilities with these techniques.   The floating-zone method would appear to have an advantage in this regard but it must be remembered that the poly-Si feed rod must also be cast or formed initially.

In the traditional process, the Czochralski or float-zone ingots are sawed into wafers, the front face is lapped and polished and the back face is given a polishing etch treatment.   The sawing is done with an inner-diameter saw (see Fig. 3.4a) which is circumferentially tensioned to permit a thin metal blade to be used, cutting one slice at a time. Multi-wire or multi-blade [9] saws (see Fig. 3.4b) have been demonstrated with the capability for sawing tens of wafers at one time, however.   The lapping and polishing step utilizes jigs capable of holding a number of wafers but is nevertheless slow and time-consuming.   While this adds little to the present cost in low-volume production with semiautomated equipment, it can be expected to present a major

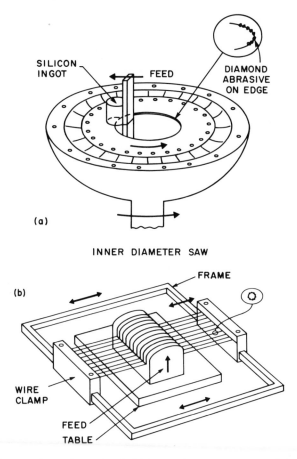

Figure 3.4 Construction schematic of (a) inner-diameter saw; and (b) multiwire saw. (Ref. 4).

bottleneck in true mass production. It should be possible to substitute a dip-etch similar to the back surface treatment for the chemical-mechanical polish, and indeed cells made from sawed and dip-etched directionally solidified Si have been shown to have good AM1 efficiencies in the 8 to 10 percent range [7].

As an alternative to the casting and sawing of ingots, direct casting of Si sheet and/or hot rolling to sheet thickness have been proposed [10]. The purification produced during directional solidification is sacrificed in the direct casting of this sheet. The reactivity of Si at the 1470C melting temperature also poses a real problem. The surface tension of molten Si is high and a nozzle casting technique would have to be used to

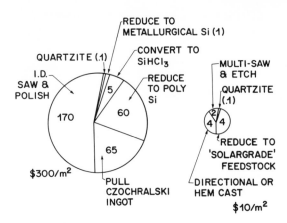

Figure 3.5   Present cost make-up and projected cost for advanced ingot process.

obtain sub-mm thicknesses. No substantial success toward successful direct thin sheet casting of Si, either single crystal or large grain semi-crystal, has been reported to date.

Rolling of thick Si sheet or ingot to thin sheet does not appear promising either. Extrusion and rolling of cm-dimension Mo-jacketed Si billets at temperatures above 1300C has been investigated but the ductility of Si does not appear to be sufficient for a practical extrusion and rolling process [11]. Rolling would almost certainly result in small-grain polycrystalline sheet and the grains would have to be regrown to larger size, for instance by float-zone regrowth using electron-beam or laser-beam heating. Highly localized heating is required to maintain a stable molten zone, since the height of the zone must be at most a few times the thickness dimension, i.e., a fraction of a mm.

The combination of an up-graded arc-furnace reduction of quartzite to provide solar-grade silicon, in combination with a directional solidification process such as the heat exchanger technique, multiple blade or wire ingot sawing, and dip-etching rather than polishing to finish the surface appears to hold the most promise at present for attaining the $10/m^2$ goal for crystalline Si wafers suitable for solar cell fabrication. Several independent reports by companies in the Si indus-try as well as by outside consulting firms have predicted that the price of semiconductor grade poly-Si can be brought down below $10/kgm [12], primarily by modification of the chloro-silane / hydrogen reduc-tion step which contributes more than 80 percent to the cost of sem-iconductor grade poly-silicon in the traditional process. The prognosis either for the quality of up-graded metallurgical silicon to be raised [13] or for the cost of chloro-silane process semiconductor grade silicon to be reduced in volume production to acceptable levels of \$5-8/kgm appears excellent. The cost increments associated with Si wafer production might then compare to traditional costs as shown in Fig. 3.5.

3.1.2 Junction Formation

The formation of a *pn* junction in or on the Si substrate may be accomplished either by diffusion or by epitaxial growth of an additional layer or layers of Si. Diffusion is the cheaper process but of course results in formation of a junction within the Si substrate, and hence requires a good quality substrate in terms of dopant level and defect density. Epitaxial growth offers the possibility of producing a relatively pure layer (buffer layer) with lower density of electrically active defects, and/or a more abrupt *pn* transition with sharper or more complex doping profiles than can be obtained with diffusion. The combination of high-quality substrates and diffused junction technology has generally been employed in the traditional approach to solar cell production.

Boron and phosphorous are the most widely used acceptor and donor species in modern Si technology. Both species are diffuse relatively slowly and provide electrical levels with low ionization energies, ~0.045 ev. They form volatile hydrides, so that diffusion may be accomplished from the gas phase with B_2H_6 or PH_3 in open tube flow systems. A typical diffusion process would involve heating to 850C in, for example, PH_3 or $POCl_3$ gas to produce an n^+p junction in a lightly boron-doped substrate. Alternatively, borax or phosphate glasses may be coated onto the wafers and diffusion accomplished in a furnace with an oxidizing atmosphere.

The dopant profiles obtained with open-tube or predeposited sources differ. In the first case the dopant concentration varies like

$$\rho(z,t) = \rho_s \mathrm{erfc}\,(z/2\sqrt{Dt}) \qquad (3.1)$$

where ρ_s is the (constant) surface concentration, Z the depth into the substrate, D the diffusion coefficient (which depends on temperature) and t the diffusion time. For a predeposited diffusant source the surface concentration decreases with time and a gaussian distribution is obtained:

$$\rho(z,t) = (2N_{os}/\sqrt{\pi Dt})\exp(-z^2/Dt) \qquad (3.2)$$

where N_{os} is the initial number of dopant atoms per unit area of substrate surface. In either case the scale length of the diffusion and hence the junction depth that will be obtained for particular background doping and surface concentrations varies as the square root of diffusion time.

The series resistance of a diffused junction Si solar cell is typically dominated by the sheet resistance of the diffused front layer. Accordingly either a deep junction (which allows the front layer to be thick) or heavy doping for high layer conductivity are desirable. With a diffused junction the average top layer doping is necessarily less than the peak level at the surface. A dead layer in which the minority carrier lifetime is much reduced can form at high doping levels; thus a good collection efficiency for carriers generated near the surface requires a thin front layer with low doping. The particular trade-off made depends on the application for which the cell is intended. Conventional N^+ on p, P diffused Si cells have junction depths of ~ 0.5 micrometer and front layer sheet resistivities in the range of tens of ohms per square. These cells have limited response to blue and violet light. Violet response is of primary interest for space applications, and advanced cells intended for AM0 illumination have been made with front layers 150 nm in thickness. A special mesh contact grid is required to reduce the series resistance to an acceptable value in this case. Since the behavior of high-quality single crystal Si cells is well understood, it is commonplace to perform computer calculations modeling cell performance for various doping profiles prior to actual diffusion.

Doping with B can also result in a dead surface layer, and the maximum solubility of B is lower than the solubility of P by a factor of about 10 ($\sim 6 \times 10^{19}$ vs $\sim 5 \times 10^{20}$ cm^{-3}). B diffused cells provide a p^+n structure which shows less degradation when exposed to the high-energy radiation encountered in the space environment. Maximum radiation resistance is presently obtained in B diffused p^+nn^+ structures which are also diffused from the back with Li and P. Lithium has a much larger diffusion constant than B or P and deep diffusions are readily obtained at low temperatures. The gradient in the lithium doping and the nn^+ junction at the back surface give rise to an electric field which increases through the base region and peaks at the back contact (see Fig. 3.6). This serves to reflect photo-generated holes toward the junction, and also is presumed to aid in drifting the Li to enhance recovery from radiation damage. The back-surface-field (BSF) design enhances current collection and minimizes the effect of a reduced diffusion length, but seems to have as its major effect an improvement of the room temperature radiation recovery time of Li diffused cells. The n^+p cells may have higher initial efficiencies but the Li diffused, p^+nn^+ BSF cells show superior over-life performance for space applications [14].

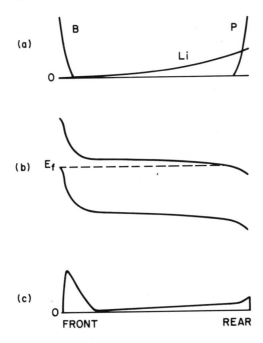

Figure 3.6 Distribution of (a) B, Li and P dopants, (b) energy band variation and (c) electric field from front to rear of an npp^+, Li diffused BSF Si solar cell.

For terrestrial applications radiation resistance or recovery is not a primary concern. Low series resistance is preferable to maximum violet response as well (compare the AM0 and AM 1.5 spectra in Figs. 1.1 and 1.2), and the simple n^+p structure appears more suitable. Additional considerations such as the cost of front contact gridding for n-compatible or p-compatible metallizations or the relative quality of low-doped n- or p-type solar grade base material will probably determine the choice between n^+p or p^+n configurations.

Homoepitaxial growth of Si on crystalline substrates is performed using the same pyrolytic reduction of $SiHCl_3$ with H_2 that is used to obtain high-purity polycrystalline Si. This is much more expensive than the simple diffusion process but allows for the creation of more complex structures. As an example, alternate p and n layers may be grown with every other junction shorted by a p^+n^+ tunnel junction. This material may be sliced edgewise to give a series-connected solar cell array with a high output voltage which facilitates impedance matching to a load [15]. Epitaxial growth also permits the crystalline quality to be improved if low quality substrate material must be used or if the very highest quality Si layers must be obtained. For that reason epitaxial layers are almost exclusively used in large-scale integrated circuit fabrication. It appears that these advantages will not prove cost-effective in the context of large-volume, minimum cost production envisaged for terrestrial solar cells. The simpler diffusion processes which routinely yield cells of 14 to 15 percent AM 1.5 efficiency should be adequate provided suitable single or semicrystalline substrate wafers indeed become available at the projected price.

3.1.3 Contacts and Antireflection Coatings

Following junction formation appropriate metal-semiconductor contacts must be established. Generally a broad-area contact is applied to the back of the cell. In some designs an effort is made to render the back contact reflective for minority carriers or long-wavelength light or both. The former is achieved with the BSF design, and the latter can be achieved to some extent with a smooth surface and a nonalloyed metal contact. The optical absorption for light just above the Si indirect band gap rises rapidly in very heavily doped or damaged Si since a large density of band-tail and in-gap states is introduced and the $k$ector conservation rule no longer strictly applies. The thickness of conventional Si cells is such that little is to be gained from an optically reflective back contact and an ohmic contact is usually utilized. For n^+p cells an evaporated Al layer, sintered at 575C, provides ohmic contact, and a back surface field can be obtained by indiffusing the Al at 800C. Very thin (50-100μm) cells have recently been prepared for space applications where weight is crucial and cost is less important; reflective backs give the current response typical of cells of standard 300μm thickness [16].

The back surface can also be tinned with a low melting point solder so that it may be bonded to an interconnecting foil or metal sheet current bus in the finished module. The choice of metal is dictated by chemical compatibility with Si and other metals present, the encapsulant materials, etc., as well as electrical conductivity and mechanical strength. Si forms a variety of intermetallic compounds which can cause resistive regions or brittle fracture and actual physical failure of contact bonds. Naturally, direct contact between such metals and Si must be avoided.

The temperature range for terrestrial applications is approximately +40 to −40C, which is less severe than the space ambient. Much of the experience on reliability of solar cell contacts from the space program should be applicable to terrestrial use. Indeed, most contact failures experienced with space cells occur prior to launch, in the terrestrial environment. The effect of humidity over extended periods undoubtedly represents the principal environmental difference. Stress caused by differential thermal expansion, thermal gradients, flexing of arrays caused by imperfect installation, wind loading, etc., together with corrosion associated with imperfect encapsulation represent the dominant failure modes for silicon cell solar modules which have undergone terrestrial environmental testing.

The front surface of the Si cell must be provided with a grid contact with substantial open area to permit light to enter the cell. Again a trade-off between obscured area and series resistance must be made, and there is no one best grid design independent of the application and environment during use. Certain principles can be followed, however. For most cases it is desirable to take the current out at one side and to provide ample area for bonding. The grid metallization should then be tapered so that the conductivity of the collector increases as the collected current increases. In principle the metallic fingers should be divided into a large number of small fingers rather than a few large ones to minimize the effect of the front layer sheet resistance (see Fig. 3.7). In practice excessively detailed metallization patterns should be avoided because of production difficulty; again a compromise must be reached in terms of the application, cell size, and other design choices which will have established the front layer sheet resistivity.

Photo-lithography is ordinarily used to define the front contact pattern. In this process a photo-sensitive polymer is applied to the surface, typically by spinning, to obtain a uniform layer. The regions where the contact is to be made are exposed through a mask to ultraviolet light and the photo-resist developed, i.e., the exposed or unexposed portions dissolved in a selective solvent. The first metal layer may then be applied by evaporation or electroplating. When the remaining photo-resist is removed the unwanted metal lifts off as well, leaving the contact pattern where the metal made contact through the previously opened windows in the photoresist. The thickness may be increased with electroless plating or by dipping into molten solder. An SiO_2 coating is sometimes used between the photo-resist and the Si surface, in which case the pattern is first opened with an HF based etch. It is possible to obtain one micrometer resolution with such techniques, but considerable saving in equipment expense results if relatively coarse featured contacts can be employed.

Further saving would result if screen-printing of conductive inks could be used directly. In this technique the contact pattern is defined in a metal screen stencil and an ink or paste consisting of metal particles and an organic binder is forced through the screen onto the cell surface. The cell is then fired to drive off the binder and sinter the metal to the cell.

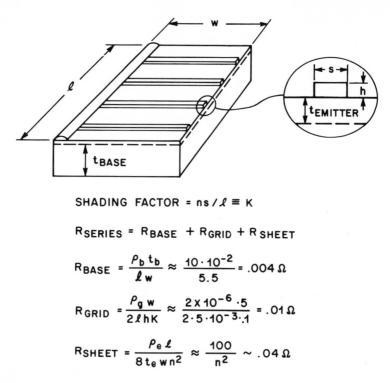

SHADING FACTOR = ns/ℓ ≡ K

R_{SERIES} = R_{BASE} + R_{GRID} + R_{SHEET}

$$R_{BASE} = \frac{\rho_b \, t_b}{\ell \, w} \approx \frac{10 \cdot 10^{-2}}{5.5} = .004 \, \Omega$$

$$R_{GRID} = \frac{\rho_g \, w}{2\ell h K} \approx \frac{2 \times 10^{-6} \cdot .5}{2 \cdot 5 \cdot 10^{-3} \cdot .1} = .01 \, \Omega$$

$$R_{SHEET} = \frac{\rho_e \, \ell}{8 \, t_e \, w \, n^2} \approx \frac{100}{n^2} \sim .04 \, \Omega$$

Figure 3.7 Contributions to series resistance for a 5 × 5 × 0.01 cm cell with Ag grid lines ($\rho_g = 10^{-6} \Omega$-cm) 10 μm thick by 10 μm wide, spaced 10 per cm on a 0.125 μm thick emitter layer of 10^{-2} Ω-cm resistivity.

An antireflection coating is necessary since the high refractive index of Si (Fig. 3.8) would otherwise result in substantial loss of power through reflection of incident light. The reflectivity at an air-semiconductor interface is given approximately by

$$R = \frac{(n_s - 1)^2}{(n_s + 1)^2} \tag{3.3}$$

where n_s is the refractive index of the semiconductor and we have ignored the imaginary part of the semiconductor dielectric constant which causes absorption. With $n_s \approx 3.5$, R would be $\sim$30%. Application of a layer of transparent dielectric with n_c in the 1.8 to 2.0 range permits substantial improvement. When the condition $n_c = \sqrt{n_s}$ is met, the reflectivity is reduced to zero for optical wavelengths satisfying

$$N_c l = \lambda/4, \ 3\lambda/4, \ 5\lambda/4, \ \text{etc.} \tag{3.4}$$

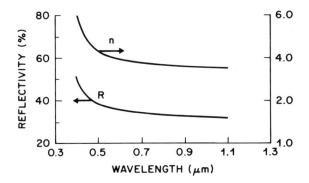

Figure 3.8 Refractive index (n) of Si and reflectivity (R) of Si/air interface as a function of optical wavelength in the visible.

where l is the thickness of the dielectric layer. The average reflectivity across the solar spectrum can be reduced below 10 percent by application of such a quarter-wavelength coating. If a material exactly satisfying $n_c = \sqrt{n_s}$ is not available one should choose a coating material with somewhat higher rather than lower value. Appropriate materials for Si are Si_3N_4 ($n_c \sim 2.0$), $Si_3O_{2x}N_{4-x}$ ($n_c \sim 1.8$ to 2.0), Al_2O_3 ($n_c = 1.86$), SiO_x ($n_c \sim 1.8$ 1.9) and TiO_2 ($n_c \sim 2.2$). Ta_2O_5 ($n_c = 2.25$) has been found to be particularly useful for the violet space cells since it does not absorb wavelengths longer than 300 nm. This is used with a quartz cover glass ($n_q = 1.5$) so that the condition $n_c \sim \sqrt{N_s N_q}$ is nearly satisfied, giving minimum reflectivity across the quartz/silicon transition.

Multiple-layer dielectric coatings offer still further improvement at the cost of complexity and manufacturing expense. For terrestrial purposes a cheap, easily applied coating is most to be desired. TiO_2 can be applied by spray coating using reaction between water vapor and tetra-isopropyl titanate at a temperature of about 150C in an air ambient [17]. Development of a process of this sort would appear essential for large scale production of cells for terrestrial use.

The cells discussed in this chapter have reasonably flat, smooth surfaces by virtue of being sawed from ingots. This facilitates application of an antireflection coating with a well-defined thickness, particularly if evaporation or sputter-deposition is used. Chemical vapor deposition tends to be uniform even on rough as well as smooth surfaces. It can be advantageous to roughen the surface of a solar cell in a controlled way so as to ensure that reflected light strikes another surface before escaping, as shown in Fig. 3.9. Such textured cells have been reported to have AM0 efficiencies of 15 percent [18].

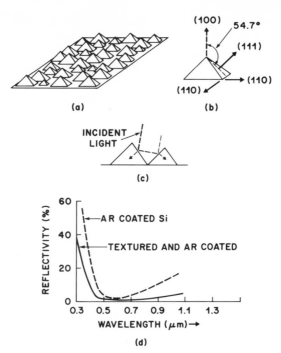

Figure 3.9 Textured or black cell principle. An anisotropic etch produces a pyramidally faceted surface (a) and (b), so that light at near normal incidence must reflect twice before escaping. The transmitted light is refracted into the Si at an angle (c) so that a thinner cell may be used as well. The reduction in reflectivity (d) and longer optical path give ~15 percent enhancement in current generation over smooth surfaced cells.

3.1.4 Si *Cells for High- or Low-level Illumination*

At illumination levels other than one sun silicon solar cells present different design problems. High or low illumination levels may imply higher or lower temperatures than usual as well, particularly in space applications involving missions traversing either much closer or much farther from the sun than the earth's orbit. Terrestrial applications at higher intensity are of importance because it is cheaper at present to build optical concentrators, using either focusing mirrors or Fresnel lenses, than solar cells. There are also low intensity terrestrial applications. Watches and calculators may employ rechargeable batteries which can be kept charged by small solar cells even in a business office environment under diffuse illumination from fluorescent lighting. The possibility of powering other office equipment, such as telephone keyboard displays, without direct connection to the office electrical supply is also under consideration.

At low illumination levels the photo-current generated is low and may become comparable to the leakage currents associated with shunt conductivity due to local junction imperfections, imperfectly passivated cell edges, etc. For this reason a highly perfect junction and maximum minority carrier lifetime is desirable for low light level cells. These features require high purity base material and state-of-the-art processes at all stages of manufacture. Fortunately series resistance becomes relatively unimportant at low light levels so that heavy doping need not be employed and a simple grid structure is sufficient.

The need for base material with minimum or zero dislocation density is even more pronounced if low temperatures are encountered simultaneously with low light levels. As the temperature is reduced the thermal velocity of minority as well as majority carrier decreases (as $\sqrt{kT}$). The effective base majority carrier concentration is reduced as well, since thermal ionization of donors and acceptors is reduced. Scattering from ionized impurities then increases as the screening provided by the majority carrier background is reduced. These effects combine to give rise to a *reduction* in the diffusion length of minority carriers in Si which is of the order of 15 percent in the best material but may be much more serious in less perfect wafers. The result is, of course, reduced current collection efficiency, seen most easily in a reduction of the short-circuit current at constant illumination as the temperature is reduced.

The effect of carrier freeze-out at low temperatures is more serious in decreasing the quality of the metal-semiconductor contacts, which tend to become rectifying as the mobile charge density is reduced and the interface depletion width increases. This can give rise to a substantial reduction in fill factor. At high intensities and higher temperatures such marginal contacts have also been found to contribute excess series resistance, so as a general rule the best possible contact quality should be pursued regardless of operating temperature or intensity.

In high temperature environments, little can be done to improve the operation of silicon solar cells. At temperatures above 350K the leakage current is determined by the intrinsic carrier concentration which varies as $\exp(E_g/kT)$. The associated drop in V_{oc} with increasing temperature more than offsets the improvement in short circuit current and the efficiency falls at a rate of $\sim 0.05\%/K$ above room temperature. The fill factor usually goes through a maximum at about 200K and also decreases with temperature above room temperature owing to thermal softening of the rectification characteristic. For terrestrial applications the temperature and insolation are not necessarily related, as external control of the cell temperature is possible. Accordingly the primary problem for high intensity terrestrial operation is reduction of the series resistance.

Silicon cells intended for operation at high insolation must be optimized in design for the intended concentration level, and will not perform as well at higher or lower levels. The optimization in principle is
a simple one of choosing base resistivity, contact pattern (and coverage
area), and junction depth simultaneously so as to obtain maximum
efficiency at the design insolation. A number of complicated designs
have been proposed and/or tested including multijunction structures
[15], and interdigitated, back surface contacted [19] cells (see Fig. 3.10)
as well as densely gridded cells of more conventional design. Recently
an efficiency exceeding 20 percent at 600 suns was attained with a
grooved-junction Si cell (see Fig. 3.11) [20]. The cell was prepared and
measured by Microwave Associates, Inc. of Burlington, Massachusetts
on contract to the U.S. Department of Energy and Sandia Laboratories.
Higher efficiencies are expected to result from refinement of this structure, e.g., by incorporating a textured front surface and moving the
junction grooves to the back as in interdigitated back contact cells.

This cell is fabricated using an anistropic chemical etch of the silicon
crystal, which permits grooves with depth-to-width ratios of greater
than 20:1 to be realized. Conventional junction diffusion and contacting are then employed. This geometry permits a substantial reduction
in series resistance without increasing the doping level in the base to

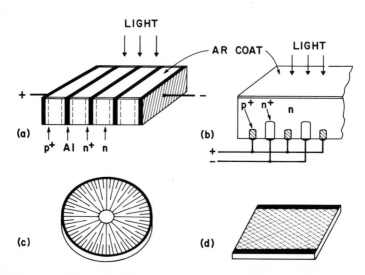

Figure 3.10   Silicon concentrator cell configurations: (a) vertical multijunction cell (cf.
Ref. 10), (b) interdigitated back contact cell to eliminate contact shadowing loss,
(c) radial fine-gridded cell for spot focus operation (Sandia Laboratories, cf. Ref. 20),
(d) redundant linear gridded cell for line focus operation (Solarex Corporation, Rockville, MD).

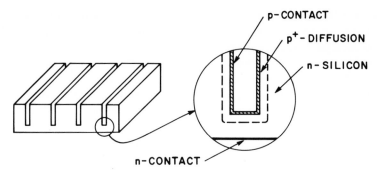

Figure 3.11   Grooved-junction silicon concentrator cell designed for low series resistance and high current collection efficiency.

the point where current collection efficiency suffers. This cell design appears particularly attractive for hybrid photovoltaic-thermal energy systems employing heat pumps, which would allow a portion of the waste heat to be utilized. As with other Si cell designs, operations at cell temperatures much above 60C does not appear to be practical, however.

3.1.5 Prognosis

Considerable experience with the operating characteristics of traditional single crystal Si cells indicates that the performance and stability of the basic diffused-junction device is more than adequate to permit fabrication of effective photovoltaic arrays. The obvious need is for cost reduction. Of the various steps involved in fabricating solar cells — refinement of starting material, ingot and wafer production, junction formation, contacting and antireflection coating — only the technology for initial quartzite reduction and junction diffusion appear clearly satisfactory at present to permit large volume production within cost goals. The low-cost silicon solar array (LSSA) portion of the U.S. National Photovoltaic Program is aimed toward reduction of the cost of the other essential steps to an acceptable level. This cost reduction is presently being achieved somewhat ahead of scheduled milestones.

Several groups have concluded that production of single or semicrystal solar grade Si wafers at a cost of less than $10/m^2$ is feasible in the near future. The problem of grid contact fabrication appears to offer a particularly difficult challenge, however. Neither photo-lithography nor screen-printing has been demonstrated to be true large-area mass production techniques. Development of a high-speed, noncontact jet printing with conductive inks appears necessary. The best performing contact at present for the n-type surface is a three-layer sandwich of Ti, Pd and Ag, which is unfortunately neither inexpensive nor amenable to screening or printing.

The primary problems in fabrication of single or semicrystal Si cells for concentrator use are reduction of series resistance and thermal sinking to restrict the cell temperature rise. Provided inexpensive optical concentrators can be manufactured, this approach appears attractive for overall system cost reduction, in view of the high efficiencies recently achieved.

3.2 Single-crystal GaAs Cells

As noted in Section 2.6, there is an optimum value for the bandgap of the semiconductor comprising the absorbing region of a solar cell. The precise value depends on the temperature of operation as well as the spectral distribution of the received radiant energy. Higher temperatures and more violet spectral distributions favor higher band-gap materials. For the region of primary terrestrial interest, i.e., T=300 to 500K and AM0 to AM2 spectra, the optimum bandgap for idealized cells lies in the 1.4 to 1.5 eV range. GaAs, with a 300K band-gap energy of 1.43 eV, provides a nearly ideal match. Theoretical one-sun maximum efficiencies calculated with measured values of minority carrier lifetime and absorption coefficients compare to Si as follows [22]: 24 percent for GaAs versus 18 percent for Si at AM0, 300K; and 28 percent versus 24 percent at AM2, 300K. At 500K, GaAs cells retain about two-thirds of the 300K efficiency, whereas Si cells are much more severely degraded to less than one quarter of the 300K value, as shown in Fig. 3.12 [23].

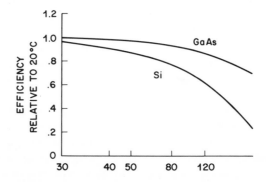

Figure 3.12   Temperature dependence of efficiency of GaAs and Si solar cells (cf. Ref. 22).

GaAs has also the advantage that it is a direct gap semiconductor so that light is absorbed near the surface. This would permit the use of thin, light-weight cells, and also means less cell volume exists to be affected by radiation damage. These features make GaAs attractive for space applications. In spite of these significant advantages, and the fact that GaAs technology is after Si the best understood and most developed of that for any semiconductor today, no GaAs solar cells have been made on a commercial basis. This is so essentially because of cost. The cost of virtually every step involved in the fabrication of a GaAs solar cell exceeds that of the Si counterpart by a large factor, as much as 100 times for substrate cost, for example. Nevertheless actual values of power conversion efficiency under terrestrial conditions in excess of 24 percent have been measured [24]. The relative insensitivity to elevated temperatures and the increased radiation resistance in comparison to Si have also been experimentally confirmed.

In special photovoltaic applications where highest possible performance is of paramount importance and cost is secondary GaAs cells represent the best present option. Single crystal GaAs cells are accordingly of primary interest in conjunction with high-intensity optical concentrators and hybrid photovaltaic-thermal energy systems, to be discussed in Chapter 5. Nontraditional (i.e., poly-crystal thin film) approaches to reduce the thickness and hence cost of GaAs cell material will be discussed in Chapter 4. Very recently the preparation of single crystal GaAs solar cells on Ge substrates has been described [25]. The quality of the GaAs epitaxial material seems similar to that prepared on GaAs substrates. Unfortunately the situation regarding cost and availability of Ge single crystal substrates, while presently better than for GaAs, is not dramatically so, and in the long range may indeed be worse.

3.2.1 Bulk GaAs Technology

Unlike Si, GaAs is a compound semiconductor and must be prepared under reactive conditions, i.e., one does not usually have a simple freezing reaction

$$\text{GaAs (liquid)} \rightarrow \text{GaAs (solid)} \qquad (3.5)$$

taking place in practice. At the melting point of GaAs (1283C) the overpressure of As vapor is 3.5 atmospheres, so that melting of GaAs in a closed vessel (SiO_2 is often used) results in a Ga rich liquid and an As gas phase. Polycrystal GaAs is formed by direct reaction of highly pure Ga and As. These elements are placed together in stoichiometric quantities in a SiO_2 ampoule which is sealed and heated slowly until reaction is complete. The polycrystal material is converted to single crystal ingot by a directional freezing (Bridgman) technique or by a modification of the Czochralski technique described earlier for Si.

In the directional freezing technique poly-crystal GaAs is placed in a SiO_2 boat with a single crystal seed at one end (see Fig 3.13). The boat is sealed in an outer SiO_2 ampoule which is placed in a furnace containing a temperature gradient, low at the seed end. The mean temperature is raised to melt the polycrystal material back just to the seed, as observed through a viewing port. The temperature is then slowly lowered over a period of days to permit the liquid to freeze out from the seed as a single crystal. Ingots of ~ 25 cm^2 in cross section and ~ 30 cm in length are prepared routinely in this way. Incorporation of dopants is something of a problem. The concentration of most dopants tends to increase in the remaining melt if the segregation coefficient differs from unity. Thus the end of the ingot which freezes last will be more heavily doped than the seed end. Depending on the specific dopant, desired doping level and end purpose, this may pose a significant problem, particularly if the bulk substrate material will form an electrically active part of the final device.

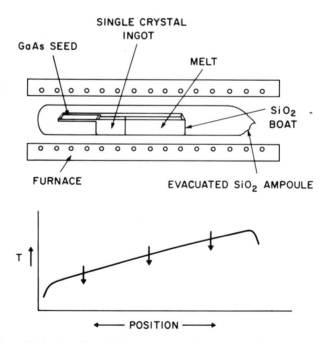

Figure 3.13   (Upper) gradient-freeze apparatus for growth of bulk single-crystal GaAs. (Lower) temperature profile in growth furnace. The temperature is uniformly decreased as a function of time, propagating the solid-melt interface to the right.

The long time at high temperature can also result in interaction between the SiO_2 boat and crucible walls and the Ga, As and dopant species. SiO is relatively volatile and reactions such as:

$$SiO_2 \rightarrow SiO_{(g)} + \tfrac{1}{2} O_{2(g)}$$

$$SiO_2 + 2Ga_{(l)} \rightarrow SiO_{(g)} + Ga_2O_{(g)} \qquad (3.6)$$

$$SiO_2 + Zn_{(g)} \rightarrow ZnO_{2(s)} + SiO_{(g)}$$

$$SiO_2 + Si_{(l)} \rightarrow SiO_{(g)}$$

all take place to some degree (Zn and Si are typical acceptor and donor dopants in this process). The result is an inevitable Si and O contamination of the ingot, so that the minimum carrier concentration in nominally undoped ingots is $n \sim 10^{16}\,cm^{-3}$. The reaction with Zn vapor during the growth of p-type ingots can be particularly troublesome and it is difficult to obtain p-type ingots with $p \leqslant 10^{18}\,cm^{-3}$.

The Czochralski growth technique may be used to pull single crystal boules of GaAs provided the arsenic overpressure at the melting point is contained. This is most successfully accomplished by encapsulating the melt with a layer of molten B_2O_3 which floats on the GaAs surface as shown in Fig. 3.14. An inert gas such as argon is introduced at sufficient pressure to prevent the loss of arsenic through the liquid B_2O_3 piston. A single crystal seed is lowered through the B_2O_3, contacted to the melt, and rotated and withdrawn as in normal Czochralski growth. Boules up to 7.5 cm in diameter have been pulled in this way. These

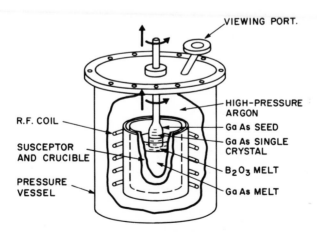

Figure 3.14   Liquid-encapsulated Czochralski growth apparatus for GaAs or InP single crystal growth.

may be rendered circular in cross-section by centerless grinding, whereas the D-shaped cross section of boat-grown Bridgman ingots is less convenient. The typical density of dislocations as indicated by surface etch to pit count is in the 2000-4000 cm^2 range for good Bridgman material, but can be much lower for liquid-encapsulated Czochralski grown GaAs. Nevertheless, no clear advantage for device fabrication (of diode lasers or light emitters as well as solar cells) has been shown to follow from use of the pulled material, and most suppliers of single-crystal GaAs today use the directional freeze technique exclusively.

GaAs is a softer and more fragile material than Si and requires greater delicacy in the slicing of ingots into wafers and the lapping and polishing of the wafers to final finish and thickness. The technology is essentially the same as for Si wafer preparation, however. Inner diameter diamond saws are used, together with mechanical polishing with alumina grits followed by chemical polishing with NaOCl in a colloidal silica solution. The minimum thickness of a 10 cm^2 wafer that can conveniently be handled without difficulty is on the order of 250 micrometers. The present cost of single crystal GaAs wafers is $3 to 4 per cm^2, roughly 100 times that of semiconducter grade Si wafers.

Ga costs less than $0.80 per gram in quantities of 10 kgm or more. As of adequate purity (six 9s or better) is somewhat less expensive, but not substantially so. Ga occurs naturally as an impurity associated with Al at about 250 ppm. It is ten times more abundant than As in the earth's crust, but the abundance in commercially exploitable ores such as bauxite is of course much lower. Ga is *not* presently recovered as an economically viable byproduct of Al manufacture by the major Al producers, owing to limited market demand. Calculations indicate that sufficient Ga would be available from Al production to permit a significant photovoltaic technology based on GaAs to develop [26], either in a thin-film context (see Chapter 4) or with single crystal cells in optical concentrators.

3.2.2 *Epitaxial Growth of* GaAs *and* $Al_xGa_{1-x}As$ *for Solar Cell Material*

Homojunction or heterojunction solar cells are fabricated on single crystal GaAs substrates by epitaxial growth of one or more layers of GaAs or $Al_xGa_{1-x}As$ alloy. The simple alternative of diffusing a dopant into the GaAs substrate wafer to form a homojunction directly has not been found to produce efficient cells. This is due to the very high surface recombination rate of minority carriers in GaAs and the short optical absorption length which dictates a high density of minority carriers at and near the surface. For blue light particularly, the response of simple homojunction GaAs cells is poor. The situation is similar to the problem of the surface dead layer in Si diffused junction cells, but is much more pronounced and does not arise from the diffusion. One way around this is to produce a very shallow, abrupt junction. This is best done by growth of a thin epitaxial layer of opposite conductivity type. If this is only a few tens of nanometers thick most of the light will be absorbed below the junction and surface recombination loss of minority carriers will be reduced to acceptable levels. A cell of this type was recently described with 17 percent AM1 efficiency, having a junction about 20 nanometers deep [27]. Alternatively, the surface may be passivated so that recombination is reduced. For GaAs the best passivant is an epitaxial layer of $Al_xGa_{1-x}As$ alloy, which must also be produced by epitaxial growth. With sufficient Al content, the top layer can serve as a transparent, conductive contact layer to a homojunction GaAs cell, or as the wide-band-gap partner of a heterojunction cell in which the *pn* junction is moved to the GaAs surface. The various configurations of GaAs-based solar cells are shown in Fig. 3.15.

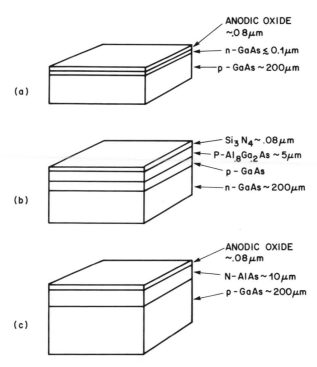

Figure 3.15 GaAs-based solar cells structures. (a) Thin n^+p homojunction structure (Ref. 26), (b) *Ppn* heteroface cell grown by LPE (cf. Ref. 21,23, 32), (c) *Np* VPE heterojunction cell (Ref. 28).

Epitaxial layers of GaAs and $Al_xGa_{1-x}As$ alloys may be obtained by growth from the liquid phase or deposition by a vapor-solid chemical reaction. Liquid phase epitaxial (LPE) techniques have seen extensive development for the production of minority carrier devices such as light emitting diodes and injection lasers while chemical vapor deposition (CVD) has been employed extensively to fabricate material for microwave devices based on majority carrier properties. GaAs solar cells were initially prepared by LPE [28] and the highest efficiencies reported so far have been achieved with that technique.

In LPE growth of GaAs a single crystal substrate is pushed under a solution of Ga saturated with As and containing small amounts of intended dopant materials. Ge and Sn are the usual acceptor and donor species; in LPE growth conditions these potentially amphoteric dopants give consistently unipolar behavior. Al is added to the solution for growth of $Al_xGa_{1-x}As$ alloy. The design of the boat and substrate slider is crucial to the quality of layers produced; a typical apparatus is shown in Fig. 3.16. Growth may occur at constant temperature if a supersaturated solution is employed, i.e., if the solution is saturated at T_1, and cooled to $T_2 < T_1$ before the substrate is pushed under the

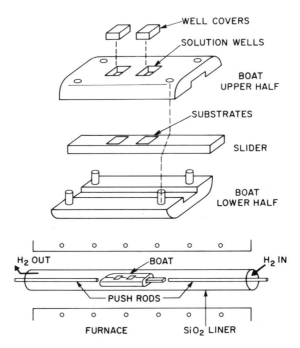

Figure 3.16 Typical apparatus for LPE growth of GaAs or $Al_xGa_{1-x}As$ layers on single crystal GaAs substrates. Upper: boat construction, lower: furnace arrangement.

well; or the furnace temperature may be slowly but continuously
decreased while the substrate is pushed under the well (and on under
subsequent wells if multilayer growth is desired).

This technique is capable of producing very high quality epitaxial
layers of $Al_xGa_{1-x}As$ for $0 \leqslant x \leqslant 0.8$. For $x \rightarrow 1.0$, the solution
becomes essentially all molten Al, which is highly reactive with trace
amounts of H_2O, CO or O_2 in the furnace ambient. The problem of
melt-back or dissolution of already grown GaAs into the high Al solution is more severe as well.

A substantial disadvantage of LPE growth as traditionally performed
is that it is intrinsically a batch process which uses new Ga solutions for
each growth run. Several gms of Ga are consumed to produce less than
one mg of epitaxial layer. The used melt may be replenished or
reclaimed by refinement and used again, but this course is not usually
followed in small scale research programs. In research-scale growth,
several cm^2 of epitaxy *per day* is about all that is produced, because of
the time required to weigh out and load the boat, to bake out and bring
the melts to thermal equilibrium, etc., although the actual growth time
may be only 10 minutes.

Various possibilities which might permit quasi-continuous LPE
growth and hence higher production rates have been suggested. A conveyor belt replacement for the slider would permit repetitive use of the
same growth solutions, although thermal cycling and/or accurate
replenishment of the melts would be required to maintain solution
saturation or super-saturation. The close mechanical tolerances found
necessary with traditional boats, including accurately reproduced substrate thickness, also argue against such a scheme. A multiple dipping
technique would avoid the close tolerances and a large melt volume
would permit many wafers to be dipped before replenishment was
required. Removal of the back-side layer could be avoided by back-to-
back double dipping. Realization of high area LPE production rates
appears difficult but perhaps not out of the question, particularly considering that these single crystal cells are considered potentially practical
only in high-ratio concentrator systems where the actual cell area is
only a small fraction of the aperture area.

Two forms of chemical vapor deposition epitaxial growth have been
used to fabricate $Al_xGa_{1-x}As$/GaAs epitaxial solar cell material. In the
first a single layer of AlAs was grown on a GaAs substrate by reaction
of AlCl with As vapor [29]. The apparatus is shown in Fig. 3.17. HCl
gas reacts with a molten Al source to form the volatile AlCl. AsH_3 is
separately introduced and decomposes pyrolytically to provide As vapor.
The AlAs is formed on a GaAs substrate held at ~1030C at a rate of

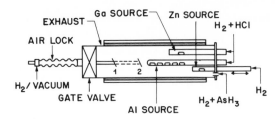

Figure 3.17   VPE growth apparatus for AlAs/GaAs growth. The corrosive nature of the AlCl intermediate requires that the tubes and boats be made from Al_2O_3 ceramic rather than the fused SiO_2 used in conventional GaAs VPE apparatus. (Reprinted with permission of the IEEE.)

up to 50 micrometers per hour, and without deliberately added dopants has $n \sim 10^{18}$ cm^{-3}. Solar cells made from material consisting of a layer of this N-AlAs grown directly on Zn-doped ($p \sim 10^{18}$ cm^3) GaAs substrates have been made with AM1 efficiencies over 20 percent. This approach offers several advantages:

1. No gallium beyond that in the substrate wafer is used during epitaxial growth.

2. Scale-up of area per growth run is easily accomplished by enlarging the reactor cross-section.

3. Several growth runs per hour may be made.

The principal drawback is that the active region of the solar cell is formed in the bulk p-GaAs substrate. The quality of cells made on substrates from different ingots has been found to vary considerably, particularly with regard to reverse current leakage and open circuit voltage. Use of AlAs rather than $Al_xGa_{1-x}As$ alloy offers maximum blue response but poses more difficult contact problems. $Al_xGa_{1-x}As$ alloy cannot be grown in this chloride transport process because the temperature ranges for good growth of AlAs and GaAs do not overlap.

A more recently exploited CVD growth process is based on the pyrolytic reaction of $(CH_3)_3$ Ga with AsH_3 on a heated GaAs substrate in a cold-wall reactor [30]. A typical apparatus is shown in Fig. 3.18. This process does permit growth of $Al_xGa_{1-x}As$ with any desired composition and with much better doping control than the AlCl process. Growth rates are typically 10 to 20 micrometers per hour, and area scaling is easily accomplished. The reaction takes place only on the heated substrate and a conveyor-belt growth apparatus based on this chemistry seems most plausible. The principal disadvantage is the high cost of $(CH_3)_3Ga$, presently $80/gm, making Ga in this form over one hun-

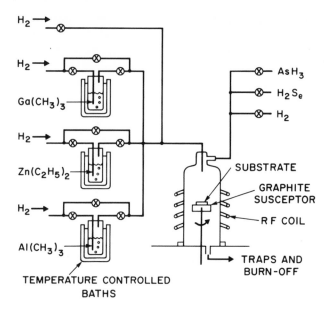

Figure 3.18 Schematic of apparatus for organometallic growth of GaAs and $Al_xGa_{1-x}As$ material.

dred times more expensive than in pure elemental form. The prospects for some cost reduction are undoubtedly good since there is only a custom specialty market for this compound at present, but the ultimate extent of such reduction has not been predicted. *Ppn* cells with excellent junction characteristics and concentrator cells with 19 percent efficiency made with organometallic sources have been reported [31]. The minority carrier lifetimes and diffusion lengths in the epitaxial GaAs layers do not appear to be quite as good as in the LPE material. This is probably the cause of the somewhat lower quantum efficiencies characteristic of these cells.

Undesired incorporation of carbon (from the methyl groups) and oxygen (from background contamination of the AsH_3 or leakage from the atmosphere) is also a problem for the $Al_xGa_{1-x}As$ layers. Unlike $(CH_3)_3Ga$, which is pyrolytically reduced in H_2 to metallic Ga and CH_4, $(CH_3)_3Al$ cracks to form a greenish-black carbide. The suboxide of Ga (Ga_2O) is volatile and its formation tends to prevent incorporation of oxygen into the GaAs layers, but the oxides of Al $(Al_2O_3, AlOOH, Al(OH)_3$ etc.) are either nonvolatile or revert to the nonvolatile Al_2O_3 form. As x increases (and values of $x \sim 0.9$ are needed for the window layer of solar cells), the incorporation of carbon and oxygen becomes much more probable. The $Al_xGa_{1-x}As$ layers thus are heavily compensated and tend to have high electrical resistance, making it difficult to achieve carrier concentrations much above the low $10^{18} cm^{-3}$ range for either p-type or n-type material. The sensitivity to trace levels of oxygen contamination poses a distinct practical problem for a manufacturing process, while the carbon problem would appear to be of a fundamental nature and may be unavoidable.

Epitaxial layers of GaAs have also been grown on GaAs and Ge single crystal substrates by reaction of GaCl and the decomposition products of AsH_3 in the vapor phase, to form thin layer n^+p homojunction cells [24]. A p^+ buffer layer is used to prevent contamination by Ge from the Ge substrates and/or to provide a back-surface field effect. Efficiencies above 21% have been reported at one sun, but the inherently high series resistance deriving from the thin top layer would seem to preclude operation at high concentration. The quality of the GaAs layers is clearly high, however, so it would seem that heterojunction cells could be prepared by pyrolysis of organometallic compounds on Ge substrates as well. The melting point of Ge (960C) is too low to permit this technique with the $AlCl$-AsH_3 growth process. Ge substrate material is presently a factor of two to three times cheaper than GaAs. It is no longer manufactured for the semiconductor industry, having been essentially completely displaced by Si, and the long-term availability is not promising.

3.2.3 Cell Fabrication

The job of antireflection coating is made easier than for Si by the fact that AlAs is transparent through much of the visible and has a nearly constant refractive index of 3.3 throughout that range. The index for low Ga $Al_xGa_{1-x}As$ alloys is of course only slightly greater as the index for GaAs is ~3.5. Little light is reflected at the AlAs GaAs interface because of the fairly close index match, which gives $R = (3.5 - 3.3)^2/(3.5 + 3.3)^2, \approx 0.1\%$. A number of dielectrics have refractive indices close to $\sqrt{3.3} = 1.8$, including the oxide glass formed directly by anodization of the AlAs [32]. Other oxides and Si_3N_4 are also useful.

AlAs and $Al_xGa_{1-x}As$ alloys with high x are hygroscopic and are readily attacked by mineral acids. The formation of a low-resistance top contact grid can pose a difficult problem, particularly for the Np configuration. P-type AlAs may be contacted with Au-2% Zn alloy, or a shallow Zn diffusion followed by a Ti-Pd-Au metallization, as used for injection laser diodes, can be employed. Each of these results in some penetration of metal toward the junction and the thermal cycling must be controlled carefully to ensure a good contact with minimum effect on junction characteristics.

It is possible to employ a top layer of GaAs for contacting purposes on either the *Np* or *Pn* devices. Oxides formed on the GaAs are much more readily removed than on AlAs and some workers have found it easier to obtain reproducibly low resistance contacts to GaAs than to AlAs [33]. The portions of the GaAs between contact grid lines are subsequently etched away by a selective etch (30% H_2O_2, pH~7.2) which stops at the AlAs layer.

Contact to *n*-type GaAs and AlAs is much less reproducibly good than to the *p*-type material where heavy Zn doping can be used. The only satisfactory contact known until recently consisted of a triple sandwich of Sn, Ni or Pd, and Au; which could be deposited in that order by evaporation or sputtering. Electrochemical deposition of 50 percent Sn-Ni alloy followed by Au plating also worked. Surprisingly, simple Au plating from very acidic (pH < 0.3) chloride baths often gives low resistance ($\sim 10^{-5}$ ohm-cm^2) contacts on GaAs, but these have poor adhesion. This last technique does not apply to AlAs, which reacts vigorously with the acidic plating bath.

The making of low -resistance ohmic contact to *n*-type GaAs remains largely art. Clearly the interface states play an important role and interfacial phases are probably of key importance. Some progress in understanding the role of these states is being made in present experiments in which the contact metals are laid down very slowly under ultra-high vacuum conditions [34]. Whether basic understanding gained in such a clean research environment can be transferred to production processes remains to be seen.

A promising new contact technology for GaAs utilizes ion implantation and subsequent laser annealing of the implanted semiconductor material [36]. In this technology, ions of the desired dopant (e.g., Te) are formed and accelerated through several hundred kilovolts. The ion source is imaged onto a separating slit by sector-focusing magnets to select only the species with desired charge/mass ratio. Additional magnets can be used to scan the ion beam over the material to be implanted. Doping densities of 10^{19} to 10^{20} cm^{-3} can be achieved in way.

Following ion implantation the semiconductor surface is left in a highly damaged and in fact usually an amorphous condition. Conventional thermal annealing requires time for heat-up and cooling, during which the dopant species may diffuse more deeply into the semiconductor than is desired and at the same time lower the high surface concentration required for contact. In compound semiconductors special encapsulation must be employed to prevent the loss of the more volatile component as well. These difficulties may be avoided by using the intense, brief heating possible with high energy pulsed lasers. Typical pulses of several joules/cm^2 lasting 10^{-7} sec cause melting of a shallow submicrometer layer which recrystallizes within 10^{-6} seconds. A very shallow region, of the order of 10 nm or less, may have to be removed from the surface to permit direct low resistance contact. Surface doping densities of 10^{21} Te atoms cm^{-3} in GaAs — over ten times the equilibrium solubility — have been achieved [35], yielding specific contact resistance of 2×10^{-5} ohm-cm^2 This value of contact resistance suggests that only about one percent of the implanted Te forms electrically active donors, however.

Se ion implantation followed by annealing with a cw Nd-YAG laser at 1.06μm wavelength has been used to fabricate n^+pp^+ solar cells. The implant is made into a vapor-phase epitaxially grown p layer. Cell efficiencies of 12 percent have been obtained [36], as compared to 17 percent for the all epitaxially grown n^+pp^+ cells [27].

The problem of low resistance contacts on GaAs cells is compounded by the economic considerations which essentially force GaAs single crystal cells to be considered only in high-concentration optical systems. At 1000 suns a current density of $\sim$25 amperes per cm^2 of cell surface must be handled. The choice of window layer (AlAs or high-Al $Al_xGa_{1-x}As$) thickness and grid pattern must be made to minimize series resistance losses. Ten amperes per cell seems a practical working level, at which a 0.4 cm^2 cell must have a series resistance of a few milliohms or less. A contact resistance of 2×10^{-5} ohm-cm^2 translates to a series resistance of about one milliohm assuming 10 percent area coverage by the grid on a 0.4 cm^2 cell. For a comb contact with 500 micrometer finger spacing the spreading resistance in a 20 micrometer thick AlAs layer ($\rho=0.02$ ohm-cm) contributes an additional 6.5 milliohms. Gold or silver grid fingers 50 micrometers wide by 10 micrometers thick contribute an additional 5 to 6 milliohms for a total of 10 to 12 milliohms. This will have a distinct effect on the cell performance since about 10 percent of the potentially available power will be dissipated in this resistance at 10 amperes output. The use of 25 micrometer wide grid lines in a cross-hatched instead of a comb array permits a reduction in the spreading resistance component, and the grid thickness may be increased somewhat by heavier electroplating to perhaps 20 microns to reduce the net series resistance to $\sim$5 milliohms. The bulk GaAs resistance and back contact resistance contribute negligibly in comparison.

Thus the window layer spreading resistance and the resistance of the metal grid dominate provided contact resistances of a few times

10^{-5}ohm-cm^2 can be achieved. This is within the state of the art for either AlAs or GaAs, n- or p-type, but the reproducibility of obtaining such high quality contacts is poor and poses a serious production problem.

3.2.4 Prognosis for Single-crystal GaAs Solar Cells

Heterojunction cells of Al$_x$Ga$_{1-x}$As or AlAs, and GaAs provide the highest efficiency, permit operation at the highest temperatures, and show least sensitivity to radiation damage of all the cell material configurations actually tested to date. There is little or no reason to expect any other material combination to offer higher performance in a single-junction device. (For a description of multijunction, spectrum splitting devices with potentially higher performance see Chapter 5. Al$_x$Ga$_{1-x}$ cells are likely to provide best performance for the larger-band-gap partner of such multijunction cells.) By the same token, the demonstrated efficiencies in the 24 percent range represent most of the theoretical ideal (26 to 27 percent) so that little room for improvement exists. Cost reduction will require reduction in the cost of GaAs single crystal substrate, reduction in the cost of Ga(CH$_3$)$_3$ for the metallorganic pyrolysis epitaxy process or development of large-volume dipping or conveyor-belt boats for liquid phase epitaxy, and development of a reliable, fine grid contacting and metallization process.

An array cost goal of \$0.50/W$_{pe}$ corresponds to a cost of \$2.00 to \$3.00 for a 0.4 cm^2 concentrator cell in a heat sink mount and implies that high-ratio optical concentrators could be built for \$50/m^2 or less, including tracking equipment. The performance advantages of GaAs cells over Si cells for space applications, particularly for missions closer to the sun than the earth's orbit, suggests that these cells will see continued development by national space agencies. While the supply of Ga and As certainly cannot compare to Si, it appears that GaAs cells could contribute significantly to terrestrial electric generation if used at high optical concentration, say greater than 500 suns. Recovery of thermal energy from the cell coolant is more beneficial than with silicon concentrator arrays because of the higher practical operating temperatures. This favors GaAs hybrid photovoltaic/thermal systems.

3.3 CdS/Cu$_2$S Cells

Solar cells made from thin film CdS/Cu$_2$S heterojunction material represent the only commercially available alternative to silicon cells at the present time. They may be produced in a variety of ways which have in common the potential for low cost, large area fabrication. The efficiencies vary from a few percent to a recently reported laboratory cell high of 9.3 percent [37], although 5 to 6 percent is a typical range for current production modules.

3.3.1 Fabrication

The construction of several CdS/Cu$_2$S cell types is shown in Fig. 3.19. An insulating substrate such as plastic, glass, or enameled steel is generally used, although conductively coated glass has been employed as well. A base electrode metallization pattern is applied by evaporation or lamination and the CdS layer formed. Vacuum evaporation or spray pyrolysis techniques have been used. In the latter technique mixed

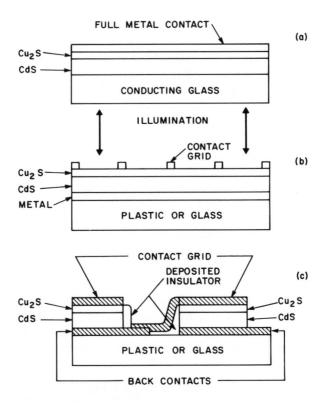

Figure 3.19   CdS/Cu$_2$S cell configurations. (a) Front-wall cell, illuminated through CdS layer, (b) back-wall cell, illuminated through Cu$_2$S layer, (c) cross-section of an integrated, series-connected contact design defined by photolithography and evaporation of insulating and metallic layers. (The top contact in (b) and (c) is in the form of a grid or striped pattern).

aqueous solutions of $CdCl_2$ and $SC(NH_2)_2$ (thiourea) are sprayed through a fog nozzle and the aerosol directed to the heated substrate at about 300C. In either case a film of about 25μm thickness with resistivity of $\sim$10 ohm-cm can be obtained, with growth rates in the range of 1 to 2 μm per minute.

The junction is formed by converting the top surface layer of CdS to Cu_2S, which may be done very simply by dipping the sheet into a hot aqueous solution of a copper salt, e.g., CuCl. As an alternative, a thin layer of Cu may be deposited and converted to sulfide by exposure to H_2S or thiourea. Following junction formation a top grid metallization is applied followed by appropriate antireflection coating and encapsulation layers.

The properties of CdS are fairly well understood, but the Cu_2S component is much less well characterized. The actual stoichiometry desired is $Cu_{2-x}S$ with $0 < x < 0.01$ corresponding to the chalcocite phase. Two other phases, djurleite and anilite occur as $x \rightarrow 0.1$. These have different crystal structure, lower optical absorption and shorter diffusion lengths. The copper deficient phases can be formed at the stage of fabrication, or during device operation, for example via formation of copper oxide if imperfect encapsulation exists. The $Cu_{2-x}S$ may also degrade (x increasing from zero) by self-anodization under forward bias:

$$Cu_2S - e^- \rightarrow xCu^+ + Cu_{2-x}S \qquad (3.7)$$

This may happen if the cell is operated at open circuit conditions at raised temperatures; the cupric ions then migrate into the grain boundaries and may cause additional problems of shunt conduction. Electrochemical reduction of Cu^+ to metallic copper can take place, in principle, for cell bias over 0.38 volts.

These instabilities of the Cu_2S layer have given CdS/Cu₂S cells the reputation for poor reliability and short life. It appears, however, that with appropriate precautions during manufacture and with complete hermetic sealing against environmental contamination CdS/Cu₂S cells can perform for at least several years without deterioration of output. The requirement for a complete hermetic seal is probably no more stringent than will be required for 20-year life of Si cell modules, and such a seal certainly will also be needed for $Al_x Ga_{1-x}As$ devices as well. Module encapsulation is required to maintain the integrity of cell interconnect metallization in any case.

The thin-film fabricating processes used for commercial cells have two drawbacks if large-scale production is contemplated. Evaporated CdS layers about 25 micrometers thick seem to be required to prevent diffusion of Cu_2S down grain boundaries to the base electrode, particularly if the dipping process is used for top layer formation. The abundance of Cd in the earth's crust is only 1 percent that of Ga, and a thinner layer would be desirable. The properties of spray-deposited layers hold promise that thinner layers can be used since there is less tendency for deep penetration of Cu through these layers than through the evaporated layers. Another difficulty is the necessity for cell series interconnection within modules, which would appear to be the most expensive step in CdS/Cu_2S module production. A proposal to overcome this with an integrated array structure [38] as shown in Fig. 3.20 may succeed in virtually eliminating the interconnect expense. This scheme is based on the fact that certain metals, such as In or Cd, short through the Cu_2S layer to the underlying CdS. By careful choice of geometry and CdS resistivity an integrated series-connected array can be achieved without the need for separating grooves between cells. The cost is paid in cell performance, however, as an additional shunt leakage path is produced. Experimental modules with packing factors of ~85 percent have been realized. The losses due to the shunting paths

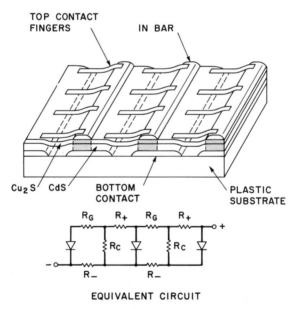

Figure 3.20 (Upper) proposed integrated, series connected contact pattern avoiding the use of photolithography with In metal junction-shorting bars (after Ref. 36). (Lower) equivalent circuit showing series resistance components from upper grid (R_G), lateral resistance of p and n layers (R_+ and R_-) and shunt resistance R_C. R_- should be as small, and R_C as large, as possible for performance approaching the standard configuration of Fig. 3.18(c).

are actually less important than series resistance losses in the top grid metallization for these cells, and it would seem that an optimized design should be able to achieve at least 7 percent module efficiency as compared to 8 percent individual cell efficiency. It remains to be seen if the In shorting stripes introduce additional degradation or failure modes, however.

3.3.2 Performance Limitations

The low efficiencies in CdS/Cu$_2$S cells result from a number of contributing factors. Cu$_2$S is an indirect gap material with fairly short diffusion length ($\sim$50nm) which dictates an optimum thickness of only about 100 nm. Since it is an indirect gap semiconductor, long wave light is not efficiently absorbed (see Fig. 3.21). CdS is a direct gap semiconductor but does not absorb light of wavelength greater than 520 nm. Hence the maximum short circuit current is less than that for GaAs cells and much less, about half, that of Si cells. The current collection efficiency is reduced by recombination at the Cu$_2$S/CdS interface which suffers from a lattice mismatch of 4.5 percent. The measured open circuit voltages are of the order of 0.5 volts as compared to a maximum of 0.8 volts expected for the Cu$_2$S energy gap. This is believed to arise from the step-type discontinuity in the conduction band at the interface shown in Fig. 3.22.

The lattice mismatch and the conduction band step can both be mitigated by alloying the CdS with ZnS. Such a film is formed by simultaneous evaporation of CdS and ZnS from two different sources. A value of 0.10 for x in the formula $Zn_xCd_{1-x}S$ has been found to increase V_{oc} from 0.5 to 0.68 volts [39], but an unexplained decrease in short circuit current (from $\sim$15 mA/cm^2 to less than 10 mA/cm^2 for one sun conditions) accompanied the improvement in voltage. The voltage improvement is presumably all due to reduction of the conduction band discontinuity as the lattice mismatch is reduced only to 3.5

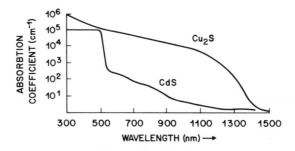

Figure 3.21   Absorption coefficients for visible light in CdS and Cu$_2$S.

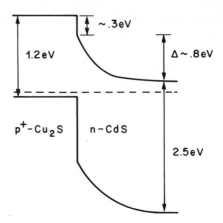

Figure 3.22 CdS/Cu$_2$S band diagram. The electron affinity of CdS is 0.3 eV greater than that of Cu$_2$S (cf. Ref. 37).

percent at $x \sim 0.1$. The resistivity of the $Zn_x Cd_{1-x}S$ increases substantially with x, giving rise to series resistance problems. The net result is that these cells have lower efficiency than the conventional CdS/Cu$_2$S cells and hence offer no advantage in practice at least in the present stage of their development. The alternative to modification of the CdS layer is alteration of the Cu$_2$S half of the cell. This requires substitution of some other p-type semiconductor. Several possibilities are under investigation at the research level and are discussed in Chapter 4, including CuInSe$_2$ and InP.

3.3.3 Prognosis

The predicted ultimate efficiency of 10 to 11% for CdS/Cu$_2$S cells is rapidly being approached in the laboratory, but it is probably unrealistic to expect efficiencies as high as this in production. There are applications where such an efficiency is acceptable provided the cost on a dollars per watt basis is low, where total power requirement is not large or area limitations are not paramount. CdS/Cu$_2$S modules have a competitive edge toward this market. The extreme simplicity of manufacture suggests that this cell structure will remain important and that improvements which unduly complicate fabrication, such as substitution of another semiconductor for the Cu$_2$S layer, will not be adopted on a commercial scale.

3.4 Summary

Three well-developed technologies exist for the fabrication of solar cells which can convert solar energy to electric power at nearly the theoretical limiting efficiency for each respective material. Each has advantages and disadvantages with regard to either cost or performance.

Single crystal Si cells have been developed for space applications and routinely provide the primary power source for satellites and missions of interplanetary exploration. The present cost (1979) of the terrestrial version of such cells, packaged into array modules of 12 to 13 percent efficiency, is about \$8.00 per peak watt (1975 dollars, large volume purchase). A reduction in price by a factor of 20 to 30 appears necessary for large-scale usage and is the goal of the United States Low-cost Silicon Solar Array (LSSA) program. Elimination of the energy-intensive trichlorosilane process presently used to purify metallurgical grade silicon is of paramount importance in meeting this goal. Casting and heat-exchanger growth of semi- or single-crystal ingot and multiple-wire sawing appear marginally satisfactory from a manufacturing cost view. The development of inexpensive contacting, interconnecting, and array sealing procedures will be necessary as well.

Single crystal silicon cells can be designed to operate at several hundred suns in optical concentrators. The principal design problem involves compromises which sacrifice low-intensity performance to achieve low series resistance and high performance at high illumination levels. The operating limitation of silicon cells in concentrators derives from the rapid fall-off of cell efficiency with temperature, a result of the 1.1 eV band-gap of Si. Concentrator cell operation above 80C does not appear to be a practical system approach with Si cells for this reason. Nevertheless the attainment in 1979 of efficiencies in the 20 percent range make actively cooled Si concentrator systems potentially practical.

Single crystal heterojunction cells of GaAs with an AlAs or $Al_xGa_{1-x}As$ top layer give the highest theoretical and experimental efficiencies for single-junction solar cells. They may be operated to at least 200C with little fall-off in efficiency because of the larger band-gap energy of GaAs (1.43 eV), and are appropriate for high-concentration applications. Such cells have been operated at intensities approaching 2000 suns and have given efficiencies over 24 percent at a few hundred suns. GaAs is much more expensive than Si and is in somewhat limited supply. Significant impact from GaAs cells on terrestrial energy needs will hinge on development and manufacture of optical concentrating arrays in the 500 to 1000 suns range at a cost of \$30.00 or less per square meter of aperture. GaAs cells have superior radiation resistance to Si and for this reason as well as for their better performance over a greater range of operating temperatures they are likely to be developed for space applications. Experience gained in the course of such development should prove of value in the eventual manufacture of cells for high-concentration terrestrial systems. Such systems are presently under experimental study but are not commercially available at this time.

The third technology offers relatively inexpensive thin-film cells of roughly half the efficiency of silicon single-crystal cells. The manufacture of CdS/Cu_2S cells or modules can be carried out by several inherently cheap methods which are compatible with mass production, i.e., vacuum evaporation or spray deposition.

The stability of these cells is adversely affected by deterioration of the Cu_2S layer. The desired chalcocite form is metastable at room temperature and is rapidly degraded by water vapor. In spite of a span of development effort extending over nearly the same length of time as silicon, no real advance in performance has resulted from improved understanding or control of the properties of the Cu_2S material. Engineering improvements in contact geometry and formation and improvement of the hermetic cell package have led to a commercially successful product with 5 to 6 percent module efficiency. A systematic program to up-grade these modules to the 10 percent efficiency goal is being sponsored by the U.S. Department of Energy (taking over earlier support by predecessor organizations), and has resulted in the announcement in 1978 of 9.3 percent laboratory efficiency.

It is possible to predict that the cost goals of $0.30 to $0.50 per peak watt could be met with each of these technologies. The characteristics of Si, GaAs and CdS cells are to a large extent complementary in that they lead naturally to segments of the photovoltaic market with different system requirements. Thus continued development and market growth for all three approaches seems probable.

REFERENCES

1. J. R. McCormick, L. D. Crossman and A. Rauchholz, *Proc. 11 PVSC,* 270 (IEEE, New York, NY 1975).

2. L. P. Hunt, V. D. Dosaj, J. R. McCormick and L. D. Crossman, *Proc. 12 PVSC,* 125 (IEEE, N.Y.,N.Y., 1975).

3. M. Wolf, H. M. Goldman and A. C. Lawson, *Proc. 13 PVSC,* 271 (IEEE, New York, NY 1978).

4. Crystal Systems, Inc. Salem, MA.

5. D. M. Mattox, *J. Vac. Sci. Tech.* **12,** 1023 (1975).

6. J. Lindmayer, *Proc. 12 PVSC,* 82 (IEEE, New York, NY, 1977).

7. H. Fischer and W. Pschunder, *Proc. 12 PVSC,* 86 (IEEE, New York, NY 1977).

8. C. P. Khattak and F. Schmid, *Proc. 13 PVSC,* 137 (IEEE New York, NY 1978).

9. T. Daud, J. K. Liu, G. A. Pollock and K. M. Koliwad, *Proc. 13 PVSC,* 142 (IEEE, New York, NY 1978).

10. M. Wolf, *J. Vac. Sci. Tech.* **12,** 984 (1975).

11. G. T. Noel, S. Kulkarni, M. Wolf, D. P. Pope and C. D. Graham, Jr., *Proc. 12 PVSC,* 168 (IEEE New York, NY 1977).

12. W. C. Breneman, E. G. Farrier and H. Morihara, *Proc. 13 PVSC,* 339 (IEEE New York, NY 1978).

13. L. P. Hunt, V. D. Dosaj, J. R. McCormick, and A. W. Rauchholz, *Proc. 13 PVSC,* 333 (IEEE New York, NY 1978).

14. A. Usami and M. Yamaguchi, *Proc. 11 PVSC,* 227 (IEEE New York, NY 1975).

15. A. Gover and P. Stella, *IEEE Trans.* **ED-21,** 351 (1974).

16. J. Lindmeyer and C. Wrigley, *Proc. 12 PVSC,* 53 (IEEE New York NY, 1977).

17. H. J. Hovel, *J. Electrochem. Soc.* **125,** 983 (1978).

18. R. A. Arndt, J. F. Allison, J. G. Haynos and A. Meulenberg, Jr., *Proc. 11 PVSC,* 40 (IEEE, New York, NY 1975).

19. M. D. Lammert, and R. J. Schwartz, *IEEE Trans.* **ED-24,** 337 (1977).

20. R. I. Frank, J. L. Goodrich and R. Kaplow *Proc. 14 PVSC,* (to be publ., IEEE, New York, NY, 1980).

21. E. L. Burgess and J. G. Fossum, *IEEE Trans.* **ED-24,** 433 (1977).

22. J. M. Woodall and H. J. Hovel, *J. Vac. Sci. Tech.* **12,** 1000 (1975).

23. L. W. James and R. L. Moon, *Proc. 11 PVSC,* 402 (IEEE New York, NY 1975).

24. R. Sahai, D. D. Edwall and J. S. Harris, *Proc. 13 PVSC,* 946 (IEEE New York, NY 1978).

25. C. O. Bozler, J. C. C. Fan and R. W. McClelland, *Proc. 7 Int. Symp. on GaAs and Related Compounds,* 429 (Inst of Phys. Conf. Series **45,** Inst. of Physics, London, UK, 1979).

26. B. L. Bryson (Private Comm., 1978).

27. J. C. C. Fan, C. O. Bozler and R. L. Chapman, *Appl. Phys. Lett.* **32,** 390 (1978).

28. J. M. Woodall and H. J. Hovel, *Appl. Phys. Lett.* **21** 379 (1972).

29. W. D. Johnston, Jr., *J. Cryst. Gr.* **39,** 117 (1977).

30. R. D. Dupuis, P. D. Dapkus, R. D. Yingling and L. A. Moudy, *Appl. Phys. Lett.* **31,** 201 (1977); H. M. Manasevit, *J. Electrochem. Soc.* **118,** 647 (1971).

31. N. J. Nelson, K. K. Johnson, R. L. Moon, H. A. Vander Plas and L. W. James, *Appl. Phys. Lett.* **33,** 26 (1978).

32. W. D. Johnston, Jr., *J. Electrochem. Soc.* **123,** 443 (1976).

33. H. A. Vander Plas, L. W. James, R. L. Moon and N. J. Nelson, *Proc. 13 PVSC,* 934 (IEEE,New York, NY 1978).

34. W. E. Spicer, P. Chye, C. M. Garner, I. Lindau, and P. Pianetta, *Surf. Science* **79,** in press.

35. P. A. Barnes, H. J. Leamy, J. M. Poate, S. D. Ferris, J. S. Williams, and G. K. Celler, *Appl. Phys. Lett.* **33,** 965 (1978).

36. J. C. C. Fan, R. L. Chapman, J. P. Donnelly, G. W. Turner, and C. O. Bozler, *Appl. Phys. Lett.* **34**, 780 (1979).

37. A. M. Barnett, J. A. Bragagnolo, R. B. Hall, J. E. Phillips, and J. D. Meakin, *Proc. 13 PVSC*, 419 (IEEE, New York, NY 1978).

38. W. J. Biter and F. A. Shirland, *Proc. 12 PVSC*, 466 (IEEE New York, NY 1977).

39. L. C. Burton, B. Baron, W. Devaney, T. L. Hench, S. Lorenz and J. D. Meakin, *Proc. 12 PVSC*, 526 (IEEE New York, NY 1977).

CHAPTER 4

Research Directions

New approaches and different photovoltaic materials may hold the promise for dramatic breakthroughs in the cost of solar cell arrays. Such unconventional approaches emphasize potentially cheap fabrication techniques and thin film geometry. In general the efficiency that has been measured is substantially less than that of conventional (i.e., single crystal) silicon or GaAs cells, and it is uncertain whether the promised low costs could in fact be met. The development effort necessary to answer the cost question will probably not be committed until performance surpassing about 10 percent AM1 efficiency can be demonstrated in the laboratory.

There are two major research directions toward breakthroughs in cost effective photovoltaics. The first is oriented toward the production of large-grain polycrystal or semicrystal Si directly in thick-film or sheet form, which would save the cost of ingot preparation and slicing. To be useful in the long run, processes of this sort must be compatible with use of refined metallurgical grade Si, although most research work at present uses semiconductor grade Si to permit isolation and evaluation of the effect of process variables. Crystalline Si sheets or films must be at least 50 μm thick for efficient light absorption. Amorphous films of silicon-hydrogen alloy behave more like direct-gap materials in that optical absorption coefficients exceeding 10^4 cm^{-1} are obtained together with a bandgap of 1.2 to 1.4 eV. Hence a few micrometers of such Si material is sufficient for thin-film solar cells.

The other principle effort is oriented toward cells based on GaAs or InP in thin film, polycrystalline form. The GaAs thin-film cells may be thought of as a potentially less expensive variant of the high performance single-crystal hetero- or homojunction cells, but most research work to date has concentrated on Schottky barrier polycrystalline cells. InP cells have been prepared with CdS or ITO (indium-tin-oxide) heterojunctions as well as in Schottky barrier form. Thin film polycrystalline GaAs material has also been evaluated as the anode in a liquid junction photoelectrochemical cell.

Finally there are several other research efforts which are further removed from established technologies. These include efforts with organic semiconductors, solid-state photothermal electric generators which use the pyroelectric rather than the photovoltaic effect, photoelectrolytic cells intended for production of H_2 under shortcircuit electrical conditions rather than operation with useful electric output, and work on photovoltaic cells fabricated from other semiconductors such as CdTe, Zn_3P_2 and the ternary chalcopyrites (e.g., $CuInSe_2$).

It is important to keep two things in mind when evaluating the promise of these research approaches. The first is that while the goal is inexpensive generation of electricity from sunlight in significant quantities, which implies compatibility with large area manufacture, demonstration of the *scientific* feasibility of reasonable power efficiency must take precedence over questions of *engineering* feasibility. Nevertheless some common sense thinking about world availability of materials is essential, particularly for the more exotic non-Si technologies. The second point is that even after (or if) scientific feasibility of 10 to 12 percent efficient cells is demonstrated, it is by no means assured that engineering feasibility will follow. Finally we must recognize that if a truly radical breakthrough indeed occurs, it is fully as likely to arise from an approach not yet described or conceived as from one of those discussed in this chapter.

4.1 Nonconventional Silicon

Slicing of single or semicrystal Si ingots is time consuming and gives rise to substantial loss of material due to kerfs and the rejection of broken or saw-damaged wafers. Silicon cannot be cast or rolled into thin sheet form of the desired 100 to 300 μm thickness without inducing a small-grain structure. There are however two variants of the Czochralski growth process which yield silicon in ribbon-like form. In the edge-defined growth technique (EFG) a refractory slotted die is used to constrain the growth cross-section [1]. A second technique inherently yields single crystal material. A ribbon-like shape is formed in a web region connecting two parallel dendrites, and no die is employed [2].

Silicon films of appropriate thickness have also been grown on ceramic substrates by dip-coating in molten silicon [3]. Graphite has been coated with metallurgical grade Si which is melted and recrystallized to yield a large-grain polycrystal substrate for subsequent epitaxial chlorosilane growth of an active device quality layer. If ribbon or sheet Si feedstock of small grain-size but otherwise satisfactory properties is produced (by one of the above methods, or by direct deposition of polysilicon onto removable substrates), travelling zone melting and regrowth using laser or conventional energy input can be performed to enhance the grain size to the point where single crystallinity is attained.

Solar cells fabricated from all these nonconventional silicon techniques are subject to the same operating principles as conventional Si single crystal cells. In particular the optical absorption characteristics and limitations of cell voltage and operating temperature arising from the Si indirect band gap value of ~ 1.1 eV characterize these cells as well. There is no hope that the cells described in this section will equal or surpass the efficiency of conventional single crystal Si cells, but there is promise for dramatic cost reduction.

4.1.1 EFG and Ribbon-to-Ribbon Silicon

The essential element in EFG growth is the die which gives the liquid-solid interface its desired cross-section [4]. The arrangement of an EFG apparatus is shown in Fig. 4.1. A single crystal seed is contacted to the melt which rises in the die channel through capillary action; withdrawal of the seed at an appropriate rate results in crystal growth with a cross-section determined by the die dimensions. Since rotation of the seed is not possible and an extended width is desired, great care must be taken to minimize and/or control transverse thermal gradients at the growth position.

Typical dimensions would be 50 to 300 μm by two to three cm for the thickness and width of the die slot which will yield ribbon of ~ 300 μm thickness and the same width as the slot. EFG ribbons up to 5 cm in width have been drawn; if the melt is replenished there is no limit to the length which could be produced in principle. Lengths of 3 to 6 meters which seem to be the largest of practical interest have been grown. The fact that the die action depends on adequate capillary attraction restricts the choice of die material somewhat, although chemical compatibility with molten Si provides a very stringent condition in any event. The die must not contain appreciable quantities of materials which are electrically active in Si, so that in practice only compounds of

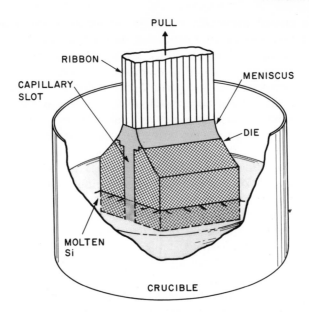

Figure 4.1   Schematic representation of an edge-defined growth apparatus with a graphite die.

C, Si, N and O are acceptable. There is insufficient capillary attraction between SiO_2 and molten Si but C and Si_3N_4 are suitable and have been used successfully. Carbon dies have the disadvantage that SiC tends to form and become incorporated as second phase inclusions in the silicon ribbon, but they are much easier to fabricate than Si_3N_4 dies and have so far given the better results.

In addition to problems caused by chemical contamination or interaction with the die, the physical shape of the die is of great importance. The meniscus between the growing crystal and the die will spread to wet the die surface to a degree determined by the contact angle between the die material and molten Si, and the height of the meniscus must be such as to give the contact angle between solid and molten Si at the growth interface. The heat of solidification must be rejected by conduction through the meniscus and die and by radiation from the crystal ribbon. The process is self-stabilizing, however, in that for a constant *linear* growth rate the ribbon thickness increases until the rates of heat of solidification and heat rejection match. This minimizes any tendency toward the thickness variations along the length dimension which occur in Czochralski growth. Residual variations in thickness along the width may be controlled by modifying the slot shape.

The maximum growth rate which can be obtained in EFG growth is determined by the rate of lateral heat removal. Since the ribbon is thin, much faster *linear* rates can be obtained than in Czochralski growth (see Fig. 4.2) as the rate varies inversely as the thickness of the ribbon (equivalent to the *radius* of the Czochralski boule). The growth rate in grams per hour, however, is similar. Unlike Czochralski growth, there is little tendency to reject impurities at the crystal-melt interface, because of the capillary die restriction which prevents the high concentration of impurities at the interface from diffusing back to the melt. When defects form in the ribbon under these conditions they are typically heavily decorated with impurities and are detrimental to the formation of high quality junctions in subsequent process steps. Because of this, highly pure Si must be used in the melt. Serious reservations must attend the ultimate practicality of EFG ribbon for that reason, at least in the as-grown form. There is a tendency for defects to propagate and perpetuate as ribbon growth proceeds, probably aided by the increasing impurity level at the melt-solid interface, and regardless of the original seed orientation a multiply-twinned structure with (110) surfaces and a (112) growth direction eventually becomes dominant. It appears that the performance of EFG ribbon could be improved considerably by zone melting and directional regrowth, since impurities could then be rejected to the molten phase in the absence of the die. This would be conceptually similar to the ribbon-to-ribbon process.

In the ribbon-to-ribbon (RTR) process [5] polycrystal feed ribbon (produced by CVD growth on a reusable graphite substrate) is zone-melted by a combination of heat sources including resistance heaters, RF induction, and/or focused laser radiation (see Fig. 4.3). This is a die-less technique and avoids the contamination problems of EFG

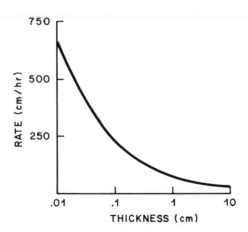

Figure 4.2   Linear growth rate for EFG Si ribbon as a function of ribbon thickness. The growth rate is essentially independent of ribbon width.

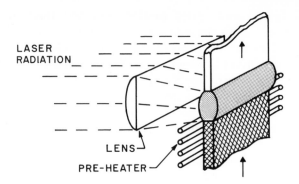

Figure 4.3   Arrangement for ribbon-to-ribbon regrowth of single crystal Si ribbon from polycrystal feed stock.   In practice both sides would be illuminated by line-focussed CO_2 lasers.

growth.   Conversion rates of polycrystal to large grain ribbon at 35 cm^2/min (5 cm width, 7 cm/min growth rate) have been reported, together with efficiencies in the 9 to 10 percent range for solar cells fabricated from RTR material.   Since the polycrystal feed stock is made by essentially the same inefficient chlorosilane reduction process now used for all semi-conductor grade polysilicon material it is doubtful that this process will be cost-effective until compatibility with an inexpensive solar-grade feed material can be shown.

At high growth rates a variety of defects appear in RTR material which are unlike those encountered in traditional ingot Si or even EFG growth.   The most dramatic of these are dendrites (oriented parallel to the ribbon) and dense clusters of stacking faults.   At lower growth rates a more gradual thermal gradient may be used and the stacking fault density can be reduced dramatically.   This process has the advantage that nearly 100 percent of the polycrystal feed stock can be converted to useful ribbon.   The expense of crucibles (a large factor in scaled-up traditional Czochralski cost estimates) and dies is also eliminated.   If a suitably cheap process for production of polycrystal in appropriately dimensioned ribbon form is realized, RTR conversion may prove attractive indeed.

4.1.2 Web-Dendrite Silicon

The dendrite-web formation [2,6] is a particular crystal structure which terminates automatically if grain boundaries originate; hence inherently single-crystal silicon with similar quality to Czochralski ingot material can be produced.   The material has the appearance of a flat strap (the web) with thickened edges where the dendritic guide-rails have grown (Fig. 4.4).   Growth is performed by introducing a small,

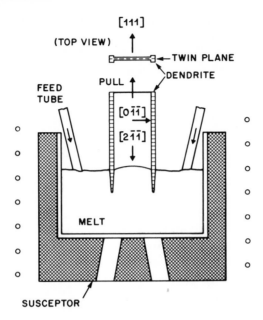

Figure 4.4 Schematic of web-dendrite growth apparatus and cross-sectional view of web-dendrite strap.

twinned dendritic seed into contact with a Si melt at thermal equilibrium (so neither premature growth nor melt-back of the seed occurs), lowering the melt temperature slightly, and withdrawing the seed slowly. Two dendrite fingers (one associated with each of the seed twins) grow out from the seed and downward into the melt when pulling starts. The web grows laterally from the dendrites. An initial taper section is used to allow slow increase of the dendrite spacing and web width, after which lengths of constant width are withdrawn. The web tends to be thicker at the center than close to the dendrites but this tendency can be minimized by careful shaping of the furnace geometry and attention to the thermal gradients near the growth interface.

Maximum linear growth rates are again determined by the rate at which latent heat can be rejected from the growth interface since nucleation occurs rapidly at the web-dendrite interface and is not a rate-limiting factor. The heat loss is geometry-determined and hence similar to the EFG process. Maximum linear growth rates of about 20m per hour can be obtained. Web thicknesses of 100 to 200 μm and widths up to 5 cm have been demonstrated. At high growth rates the rejection of impurities to the melt is less effective, just as for Czochralski growth, and high purity material must be used. Control of the furnace conditions and design of the furnace itself is critical if high yields of wide, uniform, low defect material is to be obtained, but the quality of the product is excellent and solar cells fabricated from Si web material typically have efficiencies in the 12 to 15 percent AM1 range.

Fabrication of cells from dendrite-web material is straightforward except for the difficulty presented by the lack of flatness. If the thickened dendrite regions are cut off, processing is much simpler but some material is wasted. Thus complete compatibility of this material with the existing technology for junction and contact formation cannot yet be considered well-established. The tapered leader end is useful only for smaller cells than would be desirable, but it can also be used to seed a new run. An increase in web width to ~10 cm from the present maximum of ~5 cm would also be desirable. Lengths of 20m and more have been drawn and the process is indeed compatible with continuous production. A rotationally symmetric thermal environment is neither required nor desirable for web growth since that tends to promote thickening of the web center. Thus accommodation of means for melt replenishment is easier than in the Czochralski apparatus.

4.1.3 Silicon-on-Ceramic

An alternative to the production of free-standing or self-supporting silicon sheet is growth of a suitable large grain silicon layer on a mechanically strong substrate which can double as part of the mechanical module or array structure. Several advantages are offered by an approach in which silicon is dip-coated onto a refractory ceramic substrate such as alumina or mullite (aluminum silicate) [7]. Silicon will

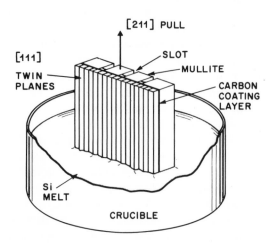

Figure 4.5   Schematic representation of SOC growth (adapted from Ref. 7, this Chapter).

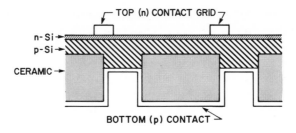

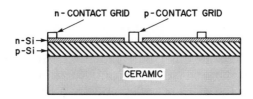

Figure 4.6 Contact arrangements for SOC solar cells: (upper), back contact made through premolded slots in ceramic substrate; (lower), interdigitated top contact pattern defined photolithographically.

not coat mullite directly, but if a carbon film is predeposited on one side of the substrate, coating is readily achieved. It is advantageous to have the silicon on only one side of the substrate, as results naturally from this approach.

Since mullite is an insulator cells may be defined on a broad-area coated slab by scribing through the silicon layer, either mechanically or with a CO_2 laser beam, and appropriate series interconnects established. Some concern must be paid to the series resistance losses which may arise in the base region, since a broad metallic base contact is not achieved as in the usual configuration. This dipping process works well on perforated mullite substrates, however, and these permit backside contacts to be made through the perforations after growth.

The dip-coating growth procedure for silicon-on-ceramic (SOC) is very simple. A mullite substrate is precoated with carbon (by pyrolytic deposition, for instance) and lowered into a crucible of molten Si (Fig. 4.5). Crystallization occurs at the top of the melt as the substate is withdrawn. Once started, growth of the film proceeds from the solid silicon and a large grain material is formed after the first cm or less of withdrawal. The quality of the film seems relatively independent of the method of carbonization of the surface of the mullite. Rubbing with high purity graphite has been found particularly effective.

This remarkably simple technique has produced crack-free silicon layers which are tightly adherent to the mullite substrate and which have a desirable columnar grain structure with the grain boundaries generally normal to the substrate plane. This results in a maximum current collection efficiency for given grain size since minority carriers can reach the junction without the need to cross grain boundaries, which are very effective recombination sites in silicon. Layers of ~100 μm thickness with 4 to 6 mm size grains have been grown at linear rates of 5 to 6 cm per minute. This is nearly an order of magnitude slower than the linear growth rates for web-dendrite growth, but the SOC process is scalable to arbitrarily large widths. Solar cells with efficiencies above 10 percent have been fabricated on SOC material of ~1.5 cm^2 area. Larger area cells have shown lower efficiencies due to reduced fill factors associated with series resistance effects, which are the primary drawback to this configuration. An interdigitated double-grid contact can be applied to the top surface as in Fig. 4.6, but excessive shading limits the efficiency which can then be obtained.

4.1.4 Silicon-On-Silicon

The contacting difficulties encountered with insulating substrates would be alleviated if Si could be grown on a metallic sheet or on graphite. The reactivity of inexpensive metals with molten Si, and the need for a close match of thermal expansion coefficients make graphite the more attractive choice. Direct deposition of silicon on graphite by reduction of chlorosilane yields small grain polycrystal material in which the grain size is invariably less than the film thickness [8]. Power efficiencies for cells made from this material are less than 2 percent.

Metallurgical grade silicon is cheap by weight (<$1/kg) but is not available in sheet form, otherwise it would be an obvious substrate choice since large grain material results from directionally solidified melts. An interesting combination substrate for CVD silicon growth is prepared by melting and directional freezing of metallurgical grade Si powder on graphite sheet [9]. The fact that Si does not wet graphite makes the substrate preparation nontrivial, but roughening of the graphite surface (or simply using a coarse saw-cut blank) inhibits break-up of the thin molten Si films.

The directional solidification is carried out as shown in Fig. 4.7. A temperature gradient is established by the combination of H_2 flow and variation of the spacing of the RF coil turns; the graphite substrate acts as its own susceptor. The temperature is raised to melt all the Si powder, then the power is reduced so that freezing takes place from the upstream end. A large grained (1 to 2 cm by several mm) film results. N^+ films are required to attain low resistance ohmic contact to the gra-

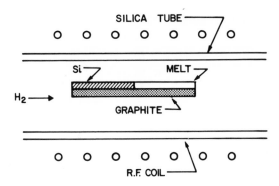

Figure 4.7   Directional solidification of Si on graphite may be accomplished continually by feeding the graphite to the left or batch-wise by reducing the RF power.

phite. This poses a need for treatment of the metallurgical grade silicon, which is typically p-type.

The metallurgical silicon can be improved by leaching with aqua regia to reduce the content of impurities which tend to segregate strongly at grain boundaries in Si, such as Fe, Cu and Al. When the cast metallurgical silicon is crushed, it breaks along grain boundaries, so that these impurities actually are concentrated at the surface of the particles in the crushed powder. Thus washing with refluxing acid is highly effective in reducing the impurity content for all but those elements with similar solubilities in solid and liquid silicon, e.g., boron or phosphorous. Following this treatment the powder may be heated with P_2O_5 to obtain n^+ doping. The surface oxide which also results is removed with hydrofluoric acid.

Epitaxial growth of n and p^+ layers of silicon on the regrown-on-graphite refined metallurgical nn^+ silicon may proceed with the usual reduction of chlorosilane in hydrogen. Cells with the structure shown in Fig. 4.8 having 9.5 percent efficiency, 9 to 10 cm^2 in area have been fabricated in this way. The principal limitation in efficiency is believed to arise from iron contamination from the metallurgical silicon layer, and from the deliberate choice of a fairly thin n-epitaxial region, 20 to 30 μm in thickness. Further improvement in refining the metallurgical grade silicon material should lead to efficiencies in the 10 to 11 percent range, but such improvements must be attained at low cost. The cost of the graphite sheet is also an open question, since graphite is presently fabricated in blocks rather than sheets, by pyrolysis of hydrocarbons at high pressure. These blocks must then be sawed to thickness, so that some of the difficulties of sliced wafer substrates remain. Because a rough surface is actually desirable and work damage is

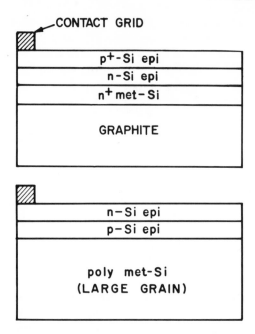

Figure 4.8   Configurations of Si on Si solar cells.

irrelevant it is not at all out of the question that the cost of sawed graphite could become low enough to meet the DOE goals, even if direct production of graphite in thin sheet form does not eventuate.

Another Si-on-Si approach involves epitaxial growth on sliced wafers of large-grain polycrystalline, refined metallurgical silicon. Some improvement of the defect structure occurs during epitaxial growth, and a maximum efficiency of 10.6 percent for a 4×4 cm^2 cell has been reported [10], in this case with less than 20 μm of epitaxially grown silicon. As discussed in Chapter 3, however, efficiencies almost as high as this have been obtained for diffused junction cells formed in large-grain polysilicon slices *without* the epitaxial growth.

It remains to be seen whether it will be less expensive to slice and coat graphite or cast and slice refined metallurgical silicon directly into substrate material. The development of a sheet-graphite fabrication process would almost certainly tip the balance to favor the directionally solidified silicon on graphite material as a substrate for chlorosilane epitaxy. Sheet graphite is also of great interest for the thin film solar cells based on InP or GaAs discussed below (Section 4.3).

4.1.5   *Crystalline* Si *Sheet on Grooved Amorphous Substrates*

When Si is deposited on an amorphous material such as fused silica sheet, an amorphous or randomly oriented fine grain material results. Subsequent recrystallization steps using heat and/or laser irradiation may be used to enhance the grain size but do little to produce the uniform grain orientation necessary for the grains to merge into single crystal material. Recently, however, it has been found that the introduction of etched grooves of appropriate geometry into the substrate prior to the deposition step permits largely single crystal material to be regrown after several laser regrowth scans [11].

The groove geometry and control of the laser power are both important. It is essential that the angle between groove wall and the top surface be appropriate to the symmetry and growth tendency of the material to be crystallized. Thus a material (such as Si) with a tendency to orient (100) planes parallel to the surface of an amorphous substrate requires grooves with an accurately square (90°) geometry. Films of materials with a tendency to orient with (111) planes parallel to the substrate would require either 70.5° or 109.5° groove angles. Either geometry can be fabricated by reactive ion etching through an appropriate mask (e.g., Cr), provided a screen grid at substrate potential is used to overcome the tendency of the ions to impact the substrate at normal incidence [12]. It is important also that the groove corners be well defined with the minimum possible radius of curvature.

Following groove etching, the mask is removed chemically and ~ 500 nm of Si is deposited by one of the conventional CVD processes, after which an argon ion laser beam in the 5 to 10 watt range is scanned in a raster pattern over the surface. The intensity must be limited so as not to damage the grating, but must also suffice to effect recrystallization. A nearly perfect degree of orientation has been achieved over several cm^2 areas, although some residual wander of orientation, on the order of 1° or so, may be observed from edge to edge. The recrystallized layers typically appear to be perfectly oriented (as determined by electron diffraction patterns) over mm^2 areas.

In the future it may be possible to fabricate grooved substrate sheets of fused silica by a molding process, although the very tight tolerances which are now required in terms of square corners and precise angular alignment will make this possiblility difficult to realize. No solar cells have yet been made on this material, and the layers would have to be grown to ~ 50 to 100μm thickness after the laser recrystallizing step in order to make solar cells. This CVD regrowth may itself be uneconomic because of the relatively thick layer required, at least in the context of $0.50/W$_{pe}$ cells for flat plate arrays.

4.2 Amorphous Si-H Alloy Cells

In the foregoing discussion of polycrystalline silicon solar cells the need for *large* grain size has been emphasized. It may indeed seem surprising that silicon material at the *opposite* extreme, that of vanishing

Figure 4.9 Schematic representations of (left) single crystal Si, amorphous Si, and hydrogenated amorphous Si. ●, Si atoms; ○, H atoms; —, bonds. The actual tetrahedral network has been represented as a square, two dimensional network for convenience.

or undefined grain size, would also be of interest. *Amorphous* silicon is not crystalline at all, yet solar cells with surprisingly good efficiencies have been made from this potentially very inexpensive material. Amorphous silicon (a-Si) films may be formed by deposition from an RF discharge established in argon, for example, between a steel substrate and a polysilicon target, or by evaporation of Si onto a substrate in vacuum. Under these conditions nominally pure a-Si films are formed, which are quite unlike crystalline Si in optical and electronic properties.

Films of *hydrogenated* a-Si can be formed from an R.F. discharge consisting of an argon-silane mixture. The degree of hydrogen incorporation is substantial (in the 10 to 40 percent range typically) and the material is properly described as an amorphous alloy. The hydrogen is incorporated as "stoppers" on what otherwise would be broken or badly distorted Si—Si bonds (Fig. 4.9), and permits a significant improvement in the electrical characteristics of the a-Si film. Hydrogenated a-Si may be prepared either p-type or n-type by adding BH_3 or PH_3 to the discharge. Schottky barrier cells of 5.5 percent efficiency have been prepared on a film of hydrogenated a-Si deposited on steel [13], and the glow discharge growth process seems readily adaptable to eventual large scale production.

4.2.1 Basic Properties of Pure and Hydrogenated a-Si

A dramatic difference between crystalline and amorphous Si is that the former has an indirect bandgap of $\sim$1.1 eV while a-Si has an optical absorption characteristic which resembles that expected for a crystal with a *direct* bandgap of $\sim$1.6 eV (Fig. 4.10). The pure a-Si material is moderately resistive ($\rho\sim10^4$ ohm-cm) and has a long wavelength optical absorption tail indicative of a high density of states within the bandgap. Much of this is removed when the material is hydrogenated, presumably as the hydrogen heals the broken Si—Si bonds.

Confirmation of the healing role of hydrogen in amorphous Si may be obtained from electron spin resonance (ESR) spectral study. Only unpaired spins have a net magnetic moment and contribute to the ESR signal; electrons in bonds are spin-paired. Hence the reduction in Si • to Si—H can be directly measured and correlated with the H content, which is in turn a function of deposition temperature and discharge parameters. In addition to the healing of broken bonds, bent, stretched, and distorted Si—Si bonds will be formed in reduced number as Si—H bonds are energetically favored.

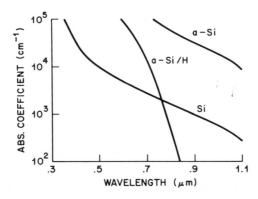

Figure 4.10   Optical absorption coefficients (typical) for amorphous and hydrogenated amorphous Si compared to single crystal Si for visible wavelengths.

In an amorphous material the translational symmetry essential to the electron band structure properties of crystals is lost. It is not possible to discuss amorphous materials in terms of an electron energy band diagram or in terms of direct or indirect energy gaps. The electron pseudo-momentum or k-vector is undefined. The electron density of states is still a function of energy, but the states are by and large localized. The nature of conduction through the electron states at a particular energy depends very much on the *spatial* density of states at that energy. The equivalent of the bandgap region in crystalline Si becomes, in amorphous Si, a region where the density of states per unit energy interval is small but nonzero. In contrast, energy states corresponding to the conduction or valence bands have high densities per unit energy interval, as shown qualitatively in Fig. 4.11. All these states are localized in the sense that they are identifiable with a particular group or cluster of atoms rather than being associated with a particular particle momentum as in the band-theoretical picture of electron states in crystalline solids.

The degree to which electrons in particular states are free to move as charge carriers depends on the degree of spatial localization of that energy state, which in turn depends on the density of states at that energy. The density of states is low, as mentioned above, in the region between the valence states and conduction states, and electron states in this gap region are highly localized. Where the electron state density is much higher, and specifically where it is above a critical density N_c, the localized states overlap and the conductivity is very much higher. Just as in a metal or semiconductor crystal, a Fermi energy may be defined by filling the total number of valence electrons (four per atom for amorphous Si) into the available states. If no donor or acceptor impurities are added, the location of the Fermi level will depend on the detailed distribution of broken, stretched and bent bonds. If E_F falls in

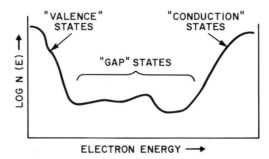

Figure 4.11   Qualitative representation of density of states as a function of electron energy for an amorphous semiconductor. The Fermi level falls within the gap region.

the gap region only hopping conductivity from one localized state to another is possible, but if E_F lies in a region of high state density the conductivity may be similar to that observed in liquid metals.

The addition of donor or acceptor impurities to an amorphous semiconductor is less effective in shifting the Fermi level than in a crystalline semiconductor. First, the impurity atom may be incorporated with a coordination number equal to its valency; for example, a boron atom may bond to three Si atoms rather than four and not form an empty orbital which acts as an acceptor. Additionally, the dopant atoms which are incorporated on four-fold coordinated sites cause only a relatively small shift of the Fermi level because of the finite density of states in the gap region as compared to the shift which would occur in a crystal in which the density of intrinsic states in the gap is strictly zero.

The addition of hydrogen to amorphous Si is beneficial in reducing the density of gap states and minimizing the probability that dopant atoms will be incorporated in other than four-fold coordinated configurations. Thus doping efficiency is enhanced, and the motion of the Fermi level for a given doping is also enhanced. The Fermi level can thus be displaced to a region of sufficiently high density of states to attain semiconducting rather than semi-insulating behavior. Reduction of the gap state density also renders the optical absorption spectrum more like that of a crystalline semiconductor. The transparency in the gap region is enhanced and the onset of the absorption edge is sharper. If *too* much hydrogen is incorporated, the probability of forming $=SiH_2$ or $-SiH_3$ groups increases and additional disorder arises. The optimum level appears to lie in the 10 to 40 percent range.

Even in the hydrogenated material one may only speak of extended range or delocalized states rather than conduction or valence band states, and localized gap states which afford only hopping mobility to the carriers. If the Fermi level falls above the energy corresponding to the critical density N_c for the conduction states (or below the energy corresponding to N_c for the valence states), behavior much like an n-type (or p-type) crystal semiconductor is observed, albeit with rather low mobility and high apparent effective mass compared to the band parameters of crystal Si. The region with N_c (conduction) $> N < N_c$ (valence) has been called a *mobility* gap. If the Fermi level could be tuned through N_c, by varying donor or acceptor density, a transition from insulating to metallic conductivity (the Anderson transition) would be observed. This can be experimentally demonstrated by varying the bias on a surface inversion layer in a MOSFET-type structure (Fig. 4.12). Detailed theoretical treatment of this transition for amorphous Si and a quantitative understanding of the localized and extended

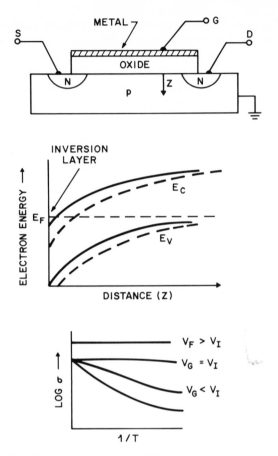

Figure 4.12 MOSFET structure (upper) in which the metal-insulator transition can be induced as the semiconductor surface layer is driven from depletion into inversion by an external bias voltage (V_G) applied to the gate G. In the metal-oxide amorphous semiconductor structure, the Fermi level can be tuned through the conductivity edge, equivalent to the condition for inversion in the single crystal case (middle). The conductivity between source and drain for such an amorphous MOSFET should be temperature independent (metallic) for gate voltages above the inversion voltage and vary as $\exp(T^{-1})$ for voltages such that the Fermi level lies below the conductivity edge (going to $\exp(T^{-1/3})$ at low temperature, cf. Ref. 12, this Chapter).

states does not yet exist. The subject of amorphous semiconductors and conducting glasses is currently a matter of substantial experimental and theoretical interest [14] and a much improved understanding of the physics and hence fundamental possibilities or limitations of these materials for photovoltaic and other applications (threshold switches, flat panel TV displays) should be forthcoming in the near future.

A recent suggestion has been made that three center bonds (Si—H—Si) as in di-borane (B_2H_6) may be formed in hydrogenated amorphous silicon [15]. These may exist with two, three, or four electrons spread over the two Si atoms and proton so that the neutral (three-electron) state can act as either an electron or a hole trap. Elimination of this center would be expected to reduce the density of gap states and improve diffusion lengths. This should be achievable by using halogenation rather than hydrogenation to heal broken bonds, since halogen atoms cannot form three center bonds. The Si—F bond length is only slightly greater than the Si—H bond length (1.56 versus 1.48Å). Fluorine addition has also been claimed to yield very much improved amorphous silicon material for solar cells [16]. It appears that gap-state density is indeed reduced by about an order of magnitude, but it is too early to conclude that this will lead to cells meeting DOE goals (no prototype results for solar cells made with fluorinated amorphous silicon have been made public as yet).

4.2.2 Fabrication and Properties of Solar Cells of Hydrogenated a-Si

Photovoltaic cells have been prepared in *pn* junction as well as Schottky barrier form on thin films of hydrogenated a-Si. Typically the a-Si is deposited on glass which has been precoated with a transparent conductor such as ITO, SnO_2, or a thin layer of chromium. N^+ doping with PH_3 added to the discharge for the initial portion of the film is used to effect ohmic contact to the transparent window. Alternatively the N^+ a-Si layer may be deposited directly on steel. (Schematic cross-sections of these various configurations are shown in Figure 4.13). The N^+ layer is followed by a layer of undoped a-Si and either a Schottky barrier metal or a p^+ a-Si layer.

Since the absorption coefficient ranges from 10^4 to 3×10^5 cm^{-1} across the visible portion of the spectrum, many of the carriers created upon illumination are created within a fraction of a micrometer of the illuminated surface. Minority carrier lifetimes in the 10^{-5} to 10^{-6} sec. range have been reported for optimum levels of H incorporation in undoped a-Si. Together with mobilities of ~2 to 10 cm^2/V·sec, maximum diffusion lengths of 1 to 2 μm can be estimated (from the relation $l^2 \sim \mu k T \tau / e$). The resistivity of this material (in the dark) is unacceptably high, however ~ 10^8 ohm-cm which would give a series resistance on the order of 10^4 ohms·cm^2 for even a 2 μm thick *i* layer *pin* device. For this reason Schottky barrier devices with a transparent metal barrier have given the best results so far, using a doping level in the a-Si active layer which compromises between good efficiency at long

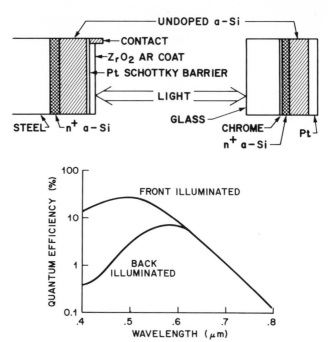

Figure 4.13   Two configurations of a-Si thin film solar cells (after Ref. 11, this Chapter). Quantum efficiency of left (front illuminated) and right (back illuminated) cell structures.

wavelength (for which a diffusion length $\gtrsim$ 1μm is needed) and reasonable resistivity. Fortunately the a-Si is photoconductive and the performance under illumination is better than would be predicted from the dark I-V characteristic alone. Analysis of these cells is complicated by this variation of conductivity with illumination level and local carrier density. Space charge and electric field effects are of major importance as determinants of cell performance.

A wide depletion layer allows the junction field to act on photogenerated carriers over a distance comparable to the optical absorption path, but the mobility is then small in the depleted region. As the carrier density is increased by doping or photoexcitation the mobility increases but the drift field region is reduced. Evidently the doping type and level, layer thicknesses, and degree of hydrogenation must be simultaneously optimized. The possibility also exists that a *pin* a-Si structure could be operated at sufficiently high concentration to ensure that the Fermi levels for both electrons and holes were driven past the critical values for the mobility gap. The high barrier voltage as compared to crystal Si would make such a-Si cells useful at elevated temperatures in photovoltaic/thermal hybrid systems.

Today, however, the best efficiency to have been reported is 5.5 percent (AM1) for an a-Si cell, using an antireflection coated Pt Schottky barrier on the illuminated surface. The active layer of undoped a-Si was about 1 μm thick in this cell, with a 5 to 10 nm thick layer of Pt and a 100 nm thick grid of Pd applied by evaporation. A ZrO_2 antireflective coating was used. The vacuum conditions during deposition of the Pt Schottky metal affect the barrier height. A small amount of O_2 in the background, for instance, leads to an AMOS structure (anti-reflection coated metal-oxide-semiconductor) with enhanced barrier height. The effect is not so pronounced as in the case of crystalline silicon/metal barriers, however. Some modification of cell properties has also been observed to occur during vacuum anneals at $\sim$300C, and some deterioration in cell properties occurs upon exposure to air at room ambient. Encapsulation will probably mitigate these instabilities since Si_3N_4 AR coated cells seem to be less affected (albeit they have not given as high an efficiency as the ZrO_2 coated cells).

Crystallization of the amorphous film would almost certainly have a drastic effect on the solar cell behavior. At least to temperatures of 300C no such tendency has been reported. Information about deleterious effects of dopant diffusion or hydrogen loss under intense illumination; high electric fields, and/or elevated temperature is also lacking. There is a marked change in the *dark I-V* characteristic of these cells after prolonged exposure to white light of $\sim$0.1 W/cm^2 intensity [17]. This appears as a two-to-three order of magnitude increase in bulk resistivity and an associated high series resistance. When illuminated, however, there is no serious increase in resistivity as the photoconductivity seems unaffected by whatever long-lived trapping effects are produced by light exposure.

Overall, hydrogenated amorphous silicon represents an exception to the observation that further progress in practical photovoltaic cells will depend more on advances in materials science and technology than on improved understanding of the underlying physics. In spite of the rather high solar conversion efficiency already obtained much basic research into the properties of amorphous silicon remains to be done. An ultimate efficiency of $\sim$15 percent has been estimated for this material [18], but that estimate is based on a number of assumptions which, while plausible, are not yet confirmed by experiment. Cells for which results have been reported so far have been typically only $\sim$0.1 to 0.2 cm^2 in size. Since there does not appear to be any reason why larger test structures could not have been fabricated one may conclude that the performance of multi-cm^2 practical-sized cells is adversely affected either by series resistance or material nonuniformity. Thus scale-up to large areas may not be so straightforward as would at first appear from a simple consideration of the technology of glow discharge film formation. Addition of fluorine certainly deserves evaluation, but this will pose severe manufacturing difficulties owing to the corrosive nature of the fluorine-containing plasma. Large area cells with 12 percent laboratory efficiency must be demonstrated before hydrogenated or halogenated amorphous silicon can be considered a serious photovoltaic option.

4.3 Polycrystalline III-V Thin Film Cells

Several members of the family of binary compound semiconductors comprised of elements in Columns IIIa and Vb of the periodic table are of special interest for solar cell research. GaAs has already been discussed in the context of single crystal cells. InP has a direct bandgap with nearly as good a match to the solar spectrum. Alloyed compounds of In, Ga, Al and As, P, and Sb may be formed, affording a range of direct band gaps across most of the visible down well into the infrared (Fig. 4.14). Other alloy compositions with indirect bandgaps afford window materials transparent through most of the visible. The fact that the lattice constant of $Al_xGa_{1-x}As$ is nearly independent of x permits

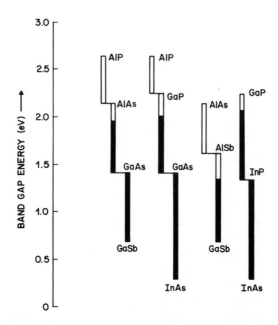

Figure 4.14 Direct (shaded) and indirect band gaps of various III-V compound semiconductors and their ternary alloys.

the fabrication of high quality heterojunctions across the entire composition range. This can only be obtained at unique compositions of other alloys, viz. $In_{.51}Ga_{.49}P/GaAs$.

All these materials are expensive and limited in availability in comparison to Si, however. In principle layers only a few micrometers thick of the direct gap compositions could be used, provided they could be formed inexpensively on a suitable cheap substrate. The choice of substrate and the growth of the semiconductor film are critical to the fabrication of efficient solar cells. The substrate should provide a strong physical bond to the semiconductor layer while at the same time making a low-resistance, nonrectifying electrical contact. It should be stable against the formation of undesired chemical phases during or after growth, and should not cause excessive doping of the semiconductor film or otherwise affect the desired electrical properties of the grown layer. A close match in thermal expansion coefficient is also necessary to avoid strain which may cause detachment when the material is cooled from the growth temperature. Metallic or at least highly conductive substrates simplify the contacting and series resistance problems.

The DOE goal of \$.50/W_{pe} corresponds to \$5.00/ft^2 of array. If one assigns an equal cost split of 25 percent to the substrate, film growth, junction formation and processing, and module assembly and encapsulation, each must come out to $\sim$\$1.00/ft^2. The choice of substrate materials is severely limited. Cold rolled steel ($\sim$\$0.23/ft^2) and soda-lime glass ($\sim$\$0.25/ft^2), in 0.5 to 1.0 mm thickness, qualify easily but neither is directly compatible with growth of III-V films.

Graphite is electrically conducting, chemically compatible with all the vapor phase film growth techniques, and can be fabricated to have a coefficient of thermal expansion between 4 and 8 $\times 10^{-6}\,K^{-1}$, so that a good match to any of the III-V semiconductors may be obtained. The cost of graphite is suitably low on a *weight* basis, but it is fabricated as blocks or bricks by high pressure pyrolysis of heavy hydrocarbons, a technique which does not seem readily adaptable to direct sheet fabrication. Thus the cost and waste of sawing operations must be considered and this will make graphite sheet too expensive to be used as substrates for flat-plate (no concentration) cells. Nevertheless most research on thin film III-V solar cells uses graphite substrates since a chemically inert substrate simplifies determination of the inherent capabilities or limitations of these polycrystalline layers.

The first thin film III-V solar cells to be reported were made by close-spaced transport of GaAs [19], and 4 to 5 percent conversion efficiencies were obtained. This growth technique requires a source of GaAs of the same area as the desired film (see Fig. 4.15). The GaAs

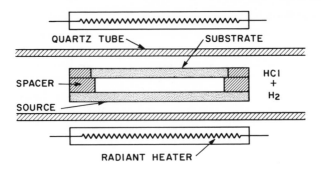

Figure 4.15 Schematic of apparatus for close-spaced growth of GaAs films.

source material is very efficiently utilized, but this technique is plainly not adaptable to mass production. The chloride-hydride (VPE) and metallorganic pyrolysis growth (MOG) techniques described earlier in Section 3.2.2 are more promising in this regard and form the basis of current research efforts. Liquid phase epitaxy has not been employed for thin film growth on foreign, noncrystalline substrates, since the metallic growth solutions generally either do not wet the substrates (as with graphite) or form undesired alloys (as with molybdenum sheet). All the binary compounds of Al, Ga and In with P, As and Sb can be grown by either VPE or MOG, although one or other method may be preferred for particular compounds. Either process offers the possibility of scaling to large area production, at least in principle.

The III-V compounds ordinarily grow in the cubic zinc-blende structure. When growth is initiated on a foreign substrate nucleation of randomly oriented grains occurs and the grains subsequently grow larger and eventually merge to form a polycrystalline film. Because of the cubic symmetry, growth normal to the substrate and along the substrate occurs at the same rate. The grain size and the film thickness at which the grains merge and the film becomes continuous are related. Naturally the semiconductor film must be continuous and pin-hole free or subsequently formed junctions will short through to the substrate. Since only films of thickness less than ~10 μm are of practical interest, grain sizes will range downward from this value. The effect of grain boundaries then becomes a very important question.

Finally, polycrystalline thin films of the III-V compounds are typically quite rough. Scanning electron micrographs of these films resemble photographs of the surface of a gravel driveway, and precise contact pattern definition using conventional photoresists is difficult or impossible. Such a textured surface is, however, beneficial from the point of view of reduced reflection. Smoother surfaces may only be attained at the cost of reduced grain size.

The structure of the polycrystalline layers depends on the subtrate material, its physical surface preparation, and the growth technique. Moderate changes in gas phase composition of the growth reactants or substrate temperature during growth have very much less effect as compared to homoepitaxial single crystal growth. The characterization of these fine-grained films is made difficult by the need to evaluate them on the conducting substrate of interest. Carrier concentration can be estimated from photoluminescence or capacitance-voltage measurements, and junction or Schottky barrier current voltage characteristics may be measured as for single crystal devices. The interpretation of these measurements is best made in comparison to similar or equivalent single crystal test structures. The latter also serve to establish a feasibility or upper limit for the solar cell efficiency that could be expected for the polycrystalline film material under ideal conditions.

4.3.1 Grain Boundary Effects in GaAs Films

Until very recently little consideration had been given, either from a theoretical or experimental viewpoint, to the nature of electron states at grain boundaries in the III-V materials. This situation is beginning to change but the discussion in this section should be regarded as tentative and as a summary of present working hypotheses rather than of clearly established concepts.

At a grain boundary the bond relations in the single crystal grain interiors can no longer be satisfied. Dangling or broken bond states are believed to be highly reactive, and just as at a free crystal-vacuum interface, the atoms will rearrange to positions where bent or stretched bonds occur but broken bonds are for the most part healed (see Fig. 4.16). The effect of this reconstruction will be to modify the density of states vs. energy distribution. This may occur in three qualitatively distinct ways (Fig. 4.17). States may be pulled out of the valence and conduction bands into the middle of the gap, into the gap edge region, or simply rearranged within the bands. The first possibility creates deep levels which will act as recombination centers and introduce depletion layers on either side of the grain boundary. The grain boundaries will then pose resistive barriers to majority carriers and act as recombination sinks for minority carriers. The second possibility resembles the band tailing which occurs at high doping levels. This will produce depletion regions, but without serious enhancement of minority carrier recombination. The third possibility only modifies the electron and hole mobilities and effective masses well into the bands and there is no major effect on minority or majority carriers. As a rough approximation, grain boundaries in Si appear to exemplify the first case, in GaAs the second, and in CdS the third.

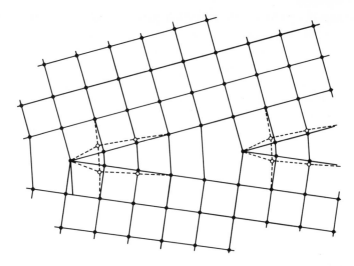

Figure 4.16 Atomic positions and bond distortions along a grain boundary (schematic). The open circles represent reconstructed atomic positions which permit a lower overall energy primarily by the relaxation of compressed bonds.

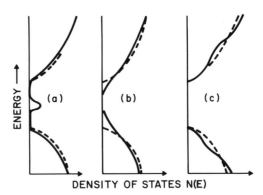

Figure 4.17 Possible effects on the density of states function due to grain boundaries in a semiconductor. The total number of states is conserved, and whether distortions occur within the gap, at the edges of the gap, or entirely within the bands depends on the details of the chemical bonding. Similar effects occur at a semiconductor surface.

If impurities are present they will tend to aggregate at grain boundaries and render the situation more complex. Because the bonds at the grain boundary region are stretched and the atoms on the average are farther apart, diffusion of impurities proceeds much more rapidly along grain boundaries than in the bulk. As a general rule diffusion processing should be avoided in a fabrication process using polycrystalline material.

The qualitative ideas behind this description of the grain boundary region may be clarified by a much-simplified analogy to molecular bonding. The potential energy of two atoms in a diatomic molecule as a function of interatomic distance is shown in Fig. 4.18 for the bonding and antibonding configurations. The potential energy of the bonding state rises much more rapidly for small decreases of interatomic distance than for small increases, hence it will be energetically favorable for atoms which are disturbed some distance from their natural bond length to form stretched, rather than compressed, bonds on average. In the stretched or dilated bond situation, the energy gap between the bonding and antibonding orbital is *reduced*. In a semiconductor crystal the conduction and valence bands correspond to the antibonding and bonding orbitals, respectively, so the expected effect at the grain boundary is a local bandgap *reduction*. The distribution of bond dilations, the effect of angular bond-orientation distortions, and the precise relation of the energy to atomic position will together determine the shape of the density of states distribution in the vicinity of the grain boundary.

Surface reconstruction studies show that significant distortions extend only to the second or third atomic layers. Hence one can expect a reduced bandgap region on the order of 2 to 3 nm in extent on either side of the grain boundary. The extent of this reduction has been estimated to be a few tenths of an eV [20]. This will give rise to a depletion region as shown in Fig. 4.19. The grain boundaries then will act as highly conductive, quasi-two-dimensional sheets within which the

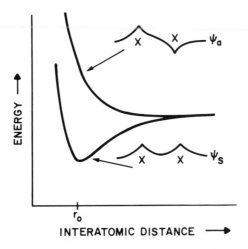

Figure 4.18   Qualitative potential energy function for two hydrogen atoms as a function of inter-atomic distance in symmetric (bonding) and antisymmetric (antibonding) states.

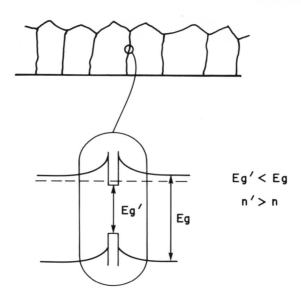

$$Eg' < Eg$$
$$n' > n$$

Figure 4.19   The effect of band gap reduction at a grain boundary is to create a degenerate, highly conductive sheet at the boundary sheathed by depletion layers extending into the crystalline grains.

carrier concentration may be very much higher than in the interior of the grains. For the case of columnar grains in Fig. 4.19, current flow normal to the substrate plane will experience essentially the bulk resistivity somewhat reduced by the enhanced grain boundary conduction. The latter effect is not large because of the ratio of boundary cross-sectional area to grain size ($\leq 10^{-3}$). Current flow *parallel* to the substrate however must cross many barriers in series. If these are of the order of 100 meV in height an apparent resistivity at room temperature many orders of magnitude greater than the true bulk resistivity can result. If the grains are not perfectly columnar and there are some horizontal grain boundaries, the current flow normal to the substrate will be somewhat impeded. Since only one or two such boundaries need be traversed the effect on a solar cell will be a moderate increase in series resistance. Thus the presence of horizontal grain boundaries will not make a *dramatic* difference, although a purely columnar structure is certainly preferable.

When a Schottky barrier or *pn* junction is formed on the polycrystalline film, one must thus consider a parallel combination of grain-face and grain-boundary diodes. The local bandgap reduction lowers the barrier height and the local concentration of carriers gives rise to a reduced depletion width for the grain boundary region as compared to the grain face region (Fig. 4.20). An equivalent circuit for a polycrystal

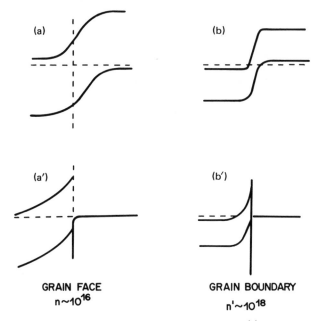

Figure 4.20   The energy band diagram of a *pn* junction lying (a) across a grain face or in a single crystal, (b) across a grain boundary. The effect on a Schottky barrier (*a'* and *b'*) can be expected to be even more severe and may lead to shorts.

Schottky diode is drawn in Fig. 4.21, where the possible presence of a linear shunt path (due to pinholes or cracks in the film, for instance) has been included.

The model of Fig. 4.21 may be checked in several ways. It predicts:

1. anomalously high resistivity parallel to the substrate;

2. normal resistivity perpendicular to the substrate for Schottky barrier or *pn* junctions;

3. an anomalously low reverse breakdown voltage, but

4. a normal *C-V* characteristic.

Normal refers to single crystal material grown under the same conditions. With metallorganic pyrolysis, growth on single crystal and graphite substrates may be carried out in the same run, with the same growth rate and hence, inferentially at least, the same doping.

The polycrystal material would be expected to show a low intensity long-wavelength photoluminescence tail as well. Various combinations of these characteristics have indeed all been observed for GaAs

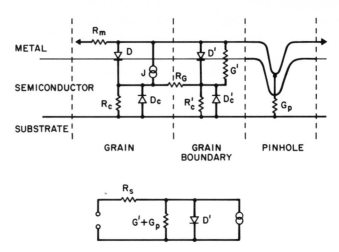

Figure 4.21   Equivalent circuit for a Schottky barrier diode made on a polycrystalline semiconductor film. If a good quality ohmic contact to the substrate is achieved, the circuit may be simplified to that shown in the lower diagram, where the dominant shunt conductance and leakage diode parameters derive from the grain boundary properties of the film. (Reprinted from Ref. 21, this Chapter, by permission of Humana Press.)

Schottky barrier devices on films prepared by MOG or chloride-hydride VPE on graphite or Mo substrates, and for AlAs/GaAs Np heterojunctions [21]. Measurements of thermally activated conductivity in MOG GaAs films on insulating substrates confirm the depletion layer model of the grain boundaries [22]. Several workers have independently concluded that the grain boundaries need not significantly affect current collection efficiency [23]. This supports the notion that mid-gap grain boundary states are absent. Recent work on *bulk* polycrystal GaAs (sliced ingot) material is also supportive of this model [24].

There are several important implications for solar cell fabrication using thin film GaAs material (or other III-V compounds, since the picture should hold for them as well). *Pn* junctions or heterojunctions should have an even larger advantage over Schottky barrier devices in the polycrystal regime than in the case of single crystal cells. The concentration of carriers at the grain boundaries appears to be in the neighborhood of $10^{18}\,\mathrm{cm}^{-3}$, a level at which *pn* junctions still have reasonably good I-V characteristics but at which Schottky barriers turn ohmic. The reduction in band gap at the grain boundary can partially be offset with a heterojunction structure. The leakage currents associated with the reduced bandgaps will be reflected in maximum open circuit voltages more to be expected from Si than from GaAs. The temperature dependence of V_{oc} and efficiency should also be more like Si.

Polycrystalline GaAs cells should have nearly the same *current* efficiency as single crystal cells, but with enhanced leakage current (J_0) and accordingly reduced V_{oc}. This is indeed observed in bulk polycrystalline Schottky barrier cells as well as heterojunction and Schottky barrier thin film cells. Further refinement and an improved theoretical basis for this model are to be desired, but it serves to provide a consistent and coherent qualitative framework for the results so far available.

4.3.2 Polycrystalline GaAs Schottky Barrier and MOS Cell Performance

The first thin film GaAs solar cells were prepared as *n*-type layers with Pt and Au Schottky barriers. Maximum solar conversion efficiencies were in the 4 to 5 percent range. This performance level has only recently been surpassed, and the best currently reported conversion efficiency for polycrystalline GaAs thin film cells is 6.3 percent [25]. The discussion in the preceding section would suggest this performance may be about all that can be expected as a result of the band-gap reduction at grain boundaries. Originally many research workers expected Schottky barrier cells to offer better performance because of the absence of the anticipated problems of *pn* junction formation, however, and the great bulk of polycrystalline GaAs cell work has been on Schottky barrier structures.

Intimate Schottky barrier contacts to GaAs result in moderate open circuit voltages on the order of 0.4 volts to either *p*-type or *n*-type material, and little difference is noted whether chromium, gold, platinum, or palladium is used. Deliberate oxidation of the surface prior to deposition of the Schottky metal has proved beneficial for single crystal GaAs Schottky barrier cells. Provided the oxide thickness is limited to ~5 nm, carrier tunneling can still take place with little increase in series resistance. The oxidation procedure is optimized empiracally and the process is not understood in detail, but substantial increases of V_{oc} (to ~0.8V) have been realized on single crystal Schottky barrier cells.

Much of the basic research on these GaAs MOS cells has been carried out on single crystal substrates. Such cells should not be viewed as competitors to the *pn* heterojunction single crystal cells, but as models to provide understanding of the physics of the MOS cell under conditions which are less complex than those presented by the thin film polycrystal structures. Efficiencies up to 17 percent have been measured for suitable anti-reflection coated (AMOS) single crystal GaAs cells [26]. Typical configurations of single crystal and thin film GaAs AMOS cells are shown in Fig. 4.22.

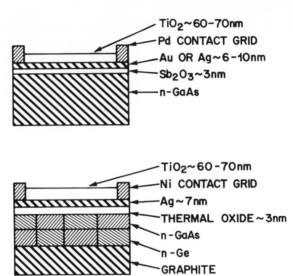

Figure 4.22   Single crystal and thin film AMOS GaAs cell structures (cf. Refs. 24 and 26, this Chapter).

The rough surface of the polycrystal GaAs films which show good quantum efficiency and hence are inferred to have adequately long minority carrier diffusion lengths poses a problem for AMOS cell fabrication. Deposition of a thin metal film with precise thickness control is not possible when the layer roughness substantially exceeds the metal film thickness and sharply angled polycrystal facets are present. The metal thickness must be increased somewhat to ensure adequate coverage, continuity, and low series resistance; but a trade-off must be made with increasing optical absorption in the metal. The effect of surface roughness on the anti-reflection coating is less severe, since the thickness of the dielectric film is ~ 80 nm and some thickness variation is offset by the refractive and multiple-bounce effects of the textured surface (compare Fig. 3.9). Thus even if the effects of grain boundaries were inconsequential, some reduction in efficiency due to the fabrication difficulties associated with a rough surface must be anticipated. The transfer of single crystal AMOS cell technology to the thin-film regime is not so straightforward as one would wish.

More important than the problem of forming the Schottky barrier, however, is the matter of growing the GaAs layer to have the right properties. Ideally a film 5 to 10 μm thick with grain size in the same range is desired. For Schottky barrier cells n-type layers with net carrier concentrations in the 10^{16} cm^{-3} range or below would give best barrier performance and minimum leakage currents. A columnar grain structure is desirable, i.e., grain boundaries should not run parallel to the substrate. Thus majority carriers may flow within the grains to the substrate without encountering depletion layer barriers at the grain boundaries. Naturally the GaAs layer should make low resistance ohmic contact to the substrate, be tightly adherent, and be free from cracks or pinholes.

Both the chloride-hydride and MOG vapor-phase techniques have been used to produce GaAs MOS cells. At this stage the MOG material is not quite so good, possibly because the grain size usually obtained is somewhat smaller than in the chloride growth. With chloride transport grain size and nucleation density (usually inversely related) may be controlled somewhat by the same techniques used to ensure ohmic contact, i.e., Zn doping for p-type and Sn or Ge doping for n-type layers, as well as by variation of the chloride mole fraction excess. The dramatic effect of an evaporated layer nominally 20 nm thick of Sn or Ge is shown in Fig. 4.23. The greatly enhanced nucleation density tends to give a grain size of 2 to 3 μm, somewhat smaller than optimum. Also shown is the effect of Zn doping, which tends to eliminate pin-holes and promote film continuity without reducing grain size. This is of interest primarily for pn junction cells, however as Schottky barriers to p-GaAs tend to have lower barrier heights than to n-type material.

With metallorganic growth no such techniques to influence grain size have been reported, although HCl addition in the gas phase may have some promise. The choice of substrate does make a difference; films grown on molybdenum tend to have stacked grains rather than the more columnar, preferred structure seen on graphite, for example. There are fewer problems in MOG with incomplete coverage and pinholes in 5 to 10 μm layers than in the case of chloride growth, and the conversion efficiency of Ga and As source material to useful films is much higher. The hybrid growth technique in which an HCl component is added to the usual MOG process would seem to offer the best of both approaches, but initial investigations showed little beneficial effect [27]. This work is all at a very early stage and it would be premature to speculate on the eventual practicality of these film growth techniques.

Another approach to achieve a controlled grain size involves recrystallization after growth. This is not practical for GaAs because of the high vapor pressure of As at the GaAs melting point. An initial layer of Ge can be grown, by pyrolysis of GeH_4 or by evaporation, and then directionally recrystallized by laser melting. GaAs will grow epitaxially on such a film since the lattice constants of GaAs and Ge are nearly equal. AMOS cells fabricated on such enhanced grain MOG GaAs-on-recrystallized-Ge films have been measured at 4.8 percent efficiency

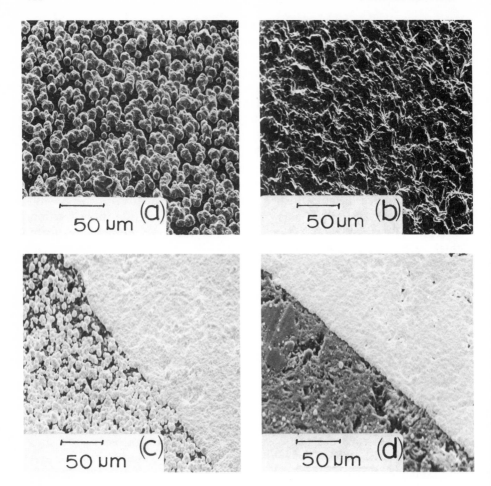

Figure 4.23 Scanning electron micrographs showing grain configuration for VPE (HCl transport) GaAs films grown under identical conditions on graphite. (a) Undoped, (b) Zn doped to give $p \sim 5 \times 10^{17}$ cm $^{-3}$, (c) upper right portion of substrate coated with nominal 20 nm thickness of evaporated Ge, undoped GaAs, (d) similar to (c) but 20 nm of Sn applied initially. The films in (c) and (d) are n-type, and low resistance ohmic contact ($\sim 10^{-4}$ ohm cm^2) resulted for films (b), (c) and (d). (Reprinted from Ref. 30, this Chapter with permission of the Electrochemical Society.)

without AR coating. This should come up to 8 percent with the standard Ta$_2$O$_5$ coating [28]. Several problems remain, however. The Ge tends to dope the GaAs p-type (n-type is desired), although this seems to depend on the surface state of the Ge film since it occurs only for particular recrystallization techniques. If tungsten is used as an interlayer to isolate steel substrates, problems with particles of tungsten on the surface of the recrystallized Ge film are encountered. Nevertheless, very thin layers of Ge and GaAs (5 and 3 μm respectively) with a relatively smooth top surface can be obtained, so that a thin deposited oxide and metal film can be applied without discontinuities. This technique is also applicable to *pn* junction cell fabrication, although in that case spreading resistance in the top layer will need to be overcome.

Recently an improvement in MOS cell efficiency has been obtained by the addition of a $GaAs_xP_{1-x}$ layer to yield an *n*-GaAs/*n*-$GaAs_xP_{1-x}$/oxide/metal structure [27]. The voltage is enhanced with little decrease in current efficiency, primarily since the surface state pinning of the Fermi level at the oxide interface occurs relative to the wider ternary band gap. The band gap of the GaAs still controls the open circuit voltage, however, but this structure does permit the MOS device to more nearly approach the performance of a GaAs *pn* junction. An efficiency before anti-reflection coating of 5 percent has been obtained for a multi-cm^2 cell, which implies 7.5 to 7.8 percent efficiency should ultimately be realizable.

4.3.3 GaAs/AlAs *Thin Film pn Heterojunctions*

A thin film version (Fig. 4.24) of the *N*-AlAs/*p*-GaAs VPE single crystal cell discussed in Section 3.2.2 has been investigated in detail [30]. The GaAs layer was grown by chloride VPE as discussed in the preceding section, with Zn doping during growth ($\sim 10^{18}$ cm^{-3}) employed to effect ohmic contact to the graphite substrate and to enhance film continuity. Layers were checked for pinholes, grain size, and thickness at this stage and an undoped AlAs ($n \sim 10^{18}$ cm^{-3}) layer was grown epitaxially over the polycrystalline GaAs.

Diodes of this material which are fully contacted over the top face (and hence of no use as solar cells) have current density vs. voltage characteristics similar to single crystal diodes of this type, although the reverse characteristic is somewhat softer. The forward current shows only a slight increase in saturation current value and a similar value of series resistance. At one sun current (~ 25 mA cm^{-2}) the forward voltage is typically well above 0.9 volts; a maximum value of 0.98 has been observed. Assuming the current generation efficiency under illumination to be high this value may be taken as an indication of V_{oc} at one sun. When a grid or stripe contact pattern is applied to the top surface it appears that only the grains *under the contact* are effective. Under one-sun illumination open circuit voltages in good agreement with the dark *J-V* prediction, i.e., exceeding 0.95 volts, have been measured. The AlAs layer however shows a very high spreading resistance, presumably due to the depletion layer barriers to majority carrier

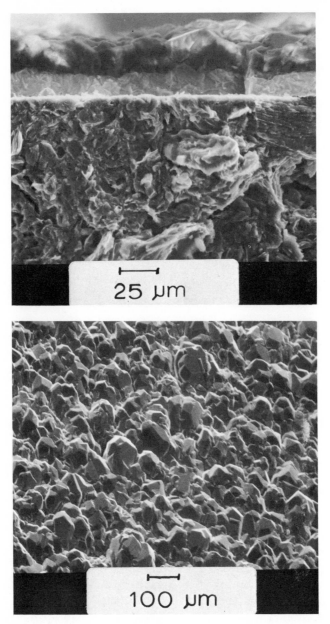

Figure 4.24 Scanning electron micrographs of (top) cross section of an *N*-AlAs/*p*-GaAs/graphite thin film heterodiode and (bottom) view of upper (AlAs) surface. (Both views at 45° perspective, the magnification anisotropy has been corrected electronically). (Reprinted from Ref. 28, Ch. 3 with permission of the North Holland Publishing Co.)

flow. Indeed, the lateral resistivity is in excess of 10^5 ohm·cm, as compared to $\sim 10^{-2}$ ohm·cm for current flow across the *pn* junction.

Because of this high lateral resistivity, the photo-current efficiency for these diodes is not a well-defined quantity. High efficiencies (>90 percent) are obtained for light introduced next to and under a contact stripe, but no meaningful value for a solar cell geometry has been obtained. Efforts to add a transparent top layer of highly conducting ITO or Cd_2SnO_4 have not been successful to date because of formation of an insulating Al_2O_3 interphase layer. Values of the fill factor and V_{oc} inferred from the dark *J-V* curves, and internal quantum efficiency calculated from the shape of the photocurrent spectral response, are essentially the same as for medium-quality single crystal cells of 17 percent AM 1.5 efficiency.

The problem of high lateral resistance to majority carrier flow is also common to *pn* homojunction thin film cells. A thin film version of the thin n^+p GaAs homojunction cell may prove easier to overcoat with ITO to obtain an effective transparent contact to all the grains, since one will not have to contend with Al_2O_3 formation.

4.3.4 CdS/InP and ITO/InP Thin Film Cells

Many of the problems associated with the CdS/Cu_2S cell are believed to stem from the less desirable aspects of Cu_2S. In many ways InP would form a preferable absorbing layer. On the (111) zincblende face the atomic spacing in InP forms a close match to the (0001) atomic spacing in the CdS wurtzite structure. Single crystal cells made by vapor deposition of CdS on single crystal (111) InP substrates have shown 15 percent AM1 efficiency. Several thin film forms of this structure have been evaluated.

InP may be prepared by the same chloride transport and MOG techniques used for GaAs. An undesired side reaction

$$(CH_3)_3In + PH_3 \rightarrow (CH_3)_3In{:}PH_3 \qquad (4.1)$$

takes place in the gas phase and the resulting addition compound condenses to form a black powder which is stable relative to formation of InP. Reduced reactor pressure helps, but substitution of P_2 for PH_3 (thermal cracking in an inlet furnace is convenient) is necessary to obtain high quality single crystal epitaxy [31]. It is not clear what the effect of the addition compound on polycrystal film growth may be, but it seems clear that any possibility for second-phase incorporation in the growing layer should be avoided. The chloride transport process suffers from the relative stability of InCl which necessitates high InCl/PH_3 input ratios ($\sim$40:1) and low conversion efficiency [32]. Most of the indium and phosphorous do not react and deposit as InCl and yellow or white phosphorous in the exhaust line. This necessitates frequent cleaning. This is particularly inconvenient owing to the presence of the white phosphorous, which ignites spontaneously in air.

P-type InP does not make ohmic contact to graphite unless very heavy Zn doping, which degrades barrier performance, is used. Since Zn diffuses rapidly in InP, it has not been possible to produce a p^+ layer confined to the graphite interface. The InP can be grown over a Zn doped GaAs layer, however, which at $p \sim 10^{18}\,\mathrm{cm}^{-3}$ makes an excellent intercontact. With the addition of a GaAs contact layer, the efficiency of CdS/InP polycrystal cells with graphite substrates increases from 2.8 to 5.7 percent [33]. The problem of growing the InP layer is reduced as well since growth is quasi-epitaxial on the GaAs film.

Since grain boundaries in CdS do not appear to introduce states within the bandgap no high spreading resistance effects occur and a normal contact grid procedure may be followed. These cells are stable when heated in air to several hundred Celsius, in marked contrast to the instabilities and degradation modes of the CdS/Cu$_2$S thin film cells discussed in Section 3.3.2.

InP is unique among the common semiconductors in that it forms Schottky barriers to metals in good accord with the simple theory based on electron affinities and work functions. Thus ITO forms a nearly ohmic contact to n-InP and a barrier on p-InP with blocking voltage nearly equal to the bandgap. ITO/InP single crystal cells with efficiencies up to 14.4 percent AM1 have been reported [34]. Results with polycrystal cells have been disappointing, however. It is not clear whether the poor performance is due to poor bulk properties of the InP *within* grains or to the effect of grain boundaries, but it does seem that the ITO/InP interface is more like a Schottky barrier than a *pn* junction. Since Zn is known to diffuse rapidly in InP, it may segregate at the grain boundaries and contribute to the reduced barrier performance relative to the single crystal cells. The zinc-doped grain boundaries do not seem to become effective recombination sinks for minority carriers in the case of the CdS/InP cells and there should be no difference for the ITO/InP configuration.

In addition to the lower efficiencies relative to GaAs single or polycrystal cells and the more complicated growth requiring a GaAs interface layer, there is also a question of In availability. It appears that In may be more limited in abundance than Ga in the long term, in spite of the present price relation in which In is favored by more than a factor of two. The slightly smaller bandgap of InP is not offset by increased current for terrestrial use because of the atmospheric water vapor absorption notch at 915 nm. Either ITO or CdS is much more stable than AlAs in unincapsulated form, but it now appears that terrestrial photovoltaic arrays, including those using Si cells, will have to be hermetically sealed. Thus the environmental stability of cells in unencapsulated form is not a primary consideration. Overall it is too early to draw any conclusions about which thin film cell configuration holds most long-range promise or whether GaAs or InP is clearly preferable.

4.3.5 Liquid Junction Solar Cells

A qualitatively different alternative to the usual Schottky barrier cell substitutes a liquid electrolyte for the metallic conductor. The band-bending at the semiconductor/electrolyte interface in this case results from the difference in chemical potential for electrons in the semiconductor and in the electrolyte. So far as the semiconductor is concerned, cell operation is equivalent to a metal Schottky barrier since available electron states are present across the barrier (Fig. 4.25). The electrolyte must incorporate an efficient charge transport species which is stable against undesired oxidation or reduction reactions. The selenide/polyselenide couple

$$Se_2^= + 2e^- \leftrightarrow 2Se^= \tag{4.2}$$

has been found particularly suitable for use in conjunction with an n-type GaAs anode [31]. This is operated at high pH in a solution of one molar KOH. The GaAs is subject to anodic attack in aqueous solutions unless a more readily oxidizable species is present:

$$2GaAs + 12OH^- \rightarrow Ga_2O_3 + As_2O_3 + 6H_2O + 12e^- \tag{4.3}$$

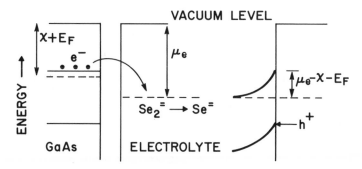

Figure 4.25 Band diagram of GaAs/selenide-polyselenide electrolyte interface. Electrons transfer to available states presented by the $Se_2^=$ ions until the space charge potential brings the Fermi level in the semiconductor into coincidence with the electron chemical potential of the electrolyte.

Since the anodic mixed oxide of GaAs is readily soluble at high and low pH, the anode corrodes. The presence of $Se^=$ suppresses this tendency since it is more easily oxidized to the polyselenide $Se_2^=$ state. The $Se_2^=$ ion diffuses to a platinum or graphite cathode where it is reduced back to $2Se^=$. It may be thought of as shuttling back and forth carrying electrons from the cathode to the GaAs anode where they combine with photogenerated holes.

The spectral response of such a cell is limited on the long-wave side by the absorption cut-off of GaAs, as usual, and on the short-wavelength side by the absorption in the polyselenide solution. The optical path length in the solution can be kept short by a Venetian blind type of geometry, for instance, which also minimizes anode-to-cathode distance and hence series resistance. With single crystal GaAs anodes efficiencies up to 9 percent have been obtained with the polyselenide solution. Lesser efficiencies resulted from CdTe or CdSe anodes or from use of the equivalent polysulfide electrolyte [35].

The surface states at the GaAs electrolyte interface are subject to *in situ* modification in this system since other electrochemical reactions may be produced by varying an externally applied bias under dark or illuminated conditions. Thus another ionic species, say of a metal which forms a stable compound with Ga and As, may be added to the electrolyte and used to stabilize or passivate the surface. Just as dilated bonds tend to introduce gap states at grain boundaries, formation of a stable monolayer compound with energetically more stable bonds may strip out gap states and push them back into the bands. This appears to happen with Ru ion treatment [36]. The efficiency of the single crystal GaAs polyselenide liquid junction cell has been increased to more than 12 percent in this way. The major factor is an improvement in open circuit voltage, although some enhancement of current efficiency is also noted, as well as the reduced photocorrosion to be expected when leakage processes are inhibited.

As with other GaAs Schottky barrier cell research, the interest in single crystal anodes is limited to establishing the feasibility of this approach and providing a basis for understanding the performance of thin film cells. Efficiencies with thin film anodes of MOG *n*-GaAs on graphite range up to 4.8 percent. Ruthenium treatment has a pronounced effect, but efficiencies without such treatment are on the order of 1 to 2 percent [37]. The current efficiency is similar to that in single crystal cells but the fill factor and open circuit voltage are both significantly lower. It appears that both tunnel leakage currents to grain boundary states and shunting currents into cracks or pin-holes are present.

In principle the liquid junction cell could be designed so that pin-hole shorts would be self-healing. If the substrate is coated with a layer of metal which forms a stable anodic oxide in the electrolyte, current through pinholes in the GaAs layer would shut off automatically when a sufficient oxide layer thickness were grown. The ruthenium treatment enhances the stability of the GaAs anode so that useful cell lifetimes of at least several years seems possible with 10 to 20 μm thick film anodes. It now appears that oxidation of the polyselenide species to elemental selenium may pose a more serious limitation to practical application of this interesting concept.

The highest thin-film-anode liquid junction cell efficiency has been obtained with rather thick GaAs layers ($\sim$40 μm) grown by chloride transport over 3 to 4 μm layers of tungsten deposited as graphite. The recently reported solar conversion efficiency of 7.3 percent following Ru ion treatment [38] represents the best efficiency for a GaAs thin film cell of any sort. Nevertheless there are very serious engineering complications to this approach. It is far too early to predict that it will turn out to be of practical value beyond the already important contribution made to our basic understanding of interfacial photochemistry.

4.4 Other Materials and Concepts

The materials discussed in the preceding three sections are all familiar, main-stream semiconductors, with the exception of a-Si, perhaps, but that is at least largely comprised of the most common semiconducting element. The question continues to arise if there might not be some little-known compound just waiting to be discovered and/or exploited for the mass production of cheap, efficient solar arrays, or even some new principle quite different from semiconductor photovoltaics which would allow cheap, efficient conversion of sunlight directly to electricity by a solid state device. Naturally not all new ideas have equal merit, and often the major benefit from exploratory work of this kind is unpredictable and even unrelated to the practical applications which motivated the initial investigations. Truly major advances in technology have occurred in the past as a result of the pursuit of ideas which went against the scientific orthodoxy of their time. Several of the ideas and new materials efforts discussed in this section have already been used to demonstrate respectable solar conversion efficiencies (i.e., 1 to 10 percent). The remainder may offer great long-range promise in spite of the fact that they show little present practicality.

4.4.1 Alternative Semiconductor Materials

In general, viable alternative photovoltaic materials to those already discussed should have a direct bandgap between 1.1 and 2 eV, be compounds of common elements, and have minority carrier diffusion lengths of several micrometers. Candidates which have been so identified are $CuInSe_2$ and several related I-III-VI$_2$ ternary chalcopyrite compounds, as well as Zn_3P_2, CdTe, and Cu_2O.

$CuInSe_2$ has a direct bandgap at 1.02 eV, and like $CuInS_2$ and $CuInTe_2$, can be doped both p-type and n-type in thin film form. The combination of p-$CuInSe_2$ and CdS is particularly favorable, giving the band diagram shown in Fig. 4.26. Solar cells have been fabricated by evaporating CdS onto single crystal $CuInSe_2$, or onto thin film $CuInSe_2$ which can be formed by a similar evaporative deposition. The former has resulted in 12 percent efficiency, while an AM1 efficiency of 6.6 percent has been obtained for the all thin film version [39]. Efficiencies in excess of 10 percent have been predicted for the thin film configuration provided improvement in the conductivity of the CdS film and better control of grain size and uniformity of the $CuInSe_2$ layer can

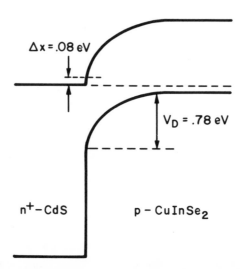

Figure 4.26   CdS/$CuInSe_2$ band-gap. Although the conduction band alignment is more favorable than for CdS/Cu_2S (compare Fig. 3.22) the energy gap of $CuInSe_2$ is lower by an offsetting amount and the maximum potential barrier heights are comparable.

be realized. Of course the long term availability of In poses a possible problem for large scale utilization of these ternary materials, just as for InP.

Cu_2O has a direct bandgap of 2.0 eV. In and Cd are useful donor dopants, and diffusion lengths of ~4 μm have been measured. In principle it can be formed very cheaply by oxidation of copper sheet, which provides a substrate of excellent thermal and conductive properties. Oxidation of an electrodeposited Cu layer on steel would result in a very inexpensive solar cell base material.

So far the best conversion efficiencies with Schottky barrier cells based on Cu_2O is only ~1 percent, as compared to theoretical estimates of ~13 percent [40]. A primary problem is the tendency for barrier metals to reduce the Cu_2O to form a barrier metal oxide and a Cu/Cu_2O junction. The resistivity of the Cu_2O is high, ~10^3 to 10^4 ohm·cm even when doped with In or Cd, if the simple grown oxide technique, which provides the principal attraction of this material, is used.

Zn_3P_2 has a direct bandgap of 1.5 eV, is normally p-type with electron diffusion lengths up to ~6 μm, and can be prepared in thin film form by evaporation of Zn_3P_2 from a heated crucible. Growth by reaction of dimethyl or diethyl zinc with phosphine would seem possible as well if the formation of addition compounds can be suppressed. Single crystal boules have been prepared by a vapor phase transport technique and a 6 percent conversion efficiency for a Mg/Zn_3P_2 single crystal Schottky barrier cell has been reported [41]. This material has the potential for equalling the performance of GaAs, and Zn and P are much more abundant (and much cheaper) than Ga and As. Intensive investigation of Zn_3P_2 has just begun and considerable work will be necessary to define the degree to which that potential can actually be realized.

CdTe is the only one of the binary II-VI semiconductors that can readily be doped to useful carrier concentrations both p-type or n-type. The bandgap is ~1.45 eV at 300K, and conductive layers may be deposited on glass substrates by evaporation, sputtering in an RF discharge, or by screen printing and firing a mixture of CdTe, $CdCl_2$, and an organic binder. The crystal structure is the cubic zincblende form, not the hexagonal wurtzite form of CdS or CdSe, but reasonably good heterojunctions have been formed by growing n-CdS on (111) oriented single crystal CdTe. These films show a strong (0001) orientation in spite of the ~10 percent lattice constant mismatch.

Two types of solar cell structures have been fabricated: n-CdS/p-CdTe heterojunction cells, and n-CdS/n-CdTe/p-CdTe homojunction cells in which the CdS layer acts as a transparent ohmic contact and may also suppress photocarrier recombination at the CdTe front face. Power conversion efficiencies of 10.5 percent have been reported for evaporated CdS on single crystal CdTe and 6.5 percent for spray-pyrolysis deposited CdS on single crystal CdTe [42]. Of most interest, however, is the value of 8.1 percent (AMO) for an *all ceramic* thin film cell made by screen printing CdS and the CdTe layers onto ITO coated glass [43]. After firing, a Cu_2Te back contact was formed and heat treated to produce a type conversion of the adjacent n-CdTe to p-type, yielding a four-layer structure of n-CdS/n-CdTe/p-CdTe/p^+-Cu_2Te. Since the pn junction is in the CdTe, potential instabilities of the Cu_2Te should be very much less significant than the analogous difficulties with the CdS/Cu_2S cell. The film thicknesses were about 10 μm and grain size in the CdTe was in the 2 to 10 μm range; a CdS film of $\sim$20 μm thickness with a resistivity of 10^{-1} ohm·cm was employed as the window contact. It would appear that the film thicknesses could be reduced to 5 μm each, which would ease the potential problem of Cd availability for large scale production, and provide some improvement in short and midwavelength current response. The ultimate theoretical efficiency of this type of cell should be about 15 percent, but the factors presently limiting efficiency are not well defined. For terrestrial illumination (AM 1.5) spectra this cell should already be close to the 10 percent efficiency goal, since the CdS window blocks an appreciable part of the blue-rich AMO spectrum used for the efficiency measurement cited above.

4.4.2 Organic Film Solar Cells

The number of possible organic compounds is many orders of magnitude greater than the number of inorganic compounds. Carbon, hydrogen, oxygen and nitrogen are abundant and cheap and the possibility that some suitable, cheap material composed primarily of them could provide a useful photovoltaic effect seems well worth investigation. To date anthracene, tetracene, phthalocyanine, chlorophyll, and hydroxy squarylium have been evaluated in some detail. Just as with inorganic compounds, a strong optical absorption band across the visible, a bandgap at about 1.5 eV, and reasonable minority and majority carrier transport properties are desired. The last area is where the organic materials have been found particularly wanting.

Most organic materials at room temperature are insulators or very poor conductors. An extensive conjugated double bond system should allow reasonable electron mobility, and this is indeed an attribute of the multiple ring compounds named above. *Pn* junctions have not been formed, and the materials show electron conductivity of 10^{-5} (ohm-cm)$^{-1}$ or less. Accordingly Schottky barrier cell configurations with thin evaporated layers of the organic material have been investigated. A metallic film intended to form an ohmic contact to the other face is also applied.

These materials all show evidence of a high density of electron traps and hence very low mobilities and short diffusion lengths, typically < 0.1 μm. This suggests that the excited electronic state is largely localized in nature, much as in inorganic amorphous semiconductors. This may arise in the organic materials by a lattice relaxation of the excited molecule, for instance. In molecular crystals the intermolecular (crystal) forces are much weaker than the intramolecular (binding) forces. Such a relaxation following electron excitation breaks the translational symmetry and creates trap states which may have long lifetimes. In addition these materials are not particularly pure by usual semiconductor standards and conventional impurity trap centers are abundant.

The high resistivities of the films primarily pose a difficulty for formation of an ohmic contact. Since the short diffusion lengths limit the useful film thickness to very small values, the series resistance added by the high bulk resistivity is not crucial. Reasonable fill factors and open circuit voltages in excess of 0.6 volts can be obtained at *low* illumination levels but the fill factors suffer as the illumination is raised to practical intensity approaching one sun. The short-circuit currents are very low, indicating poor internal quantum efficiency, and are not linearly proportional to illumination level except at very low light levels, $\leq 10^{-3}$ sun. The materials used have been found to degrade under one sun conditions. Thus efficiencies, when quoted, have generally been measured at low intensity.

The best white light efficiency reported [44] is for hydroxy-squarylium, ~ 0.1 percent at 0.01 suns and 0.02 percent at one sun (0.135 W/cm^2, AMO). It appears that use of crystalline dyes of this sort will not permit fabrication of solar cells with useful efficiency. Organic compounds with carrier diffusion lengths approaching 1 μm and suitable absorption spectra will need to be found; at present no promising candidates are known.

4.4.3 Photoelectrolytic Hydrogen Generation

The direct conversion of sunlight into electricity leaves the problem of storage of energy for nonsunlit hours, and does not directly address the problem of motor vehicle fuel. Storage batteries permit conversion of electricity into chemical energy in storable form, but with some inevitable loss in the transition and substantial investment in cumbersome storage media. Modification of the traditional photovoltaic effect can be used to generate chemical energy directly in the form of storable fuel [45]. Since we live in an oxidizing atmosphere, the stored fuel must be in the form of a suitable reducing agent: H_2, NH_3, etc. Most work has been directed toward the electrolysis of H_2O to H_2 and O_2 gas, since water is, obviously, cheap and readily available.

Hydrogen comprises a desirable fuel with many of the advantages of gasoline or natural gas. In an ordinary nonilluminated electrolytic cell a potential of several volts must be applied to achieve the electrolysis of water. Although the net free energy difference per electron transferred in the reaction

$$H_2 + \tfrac{1}{2} O_2 \rightleftharpoons H_2O \tag{4.4}$$

is only 1.23 eV, additional energy must be supplied to drive the reaction away from equilibrium, and to overcome losses at the electrode/electrolyte interfaces and in the resistance of external circuitry as well as in the electrolyte bulk. In a standard electrolysis cell with electrodes optimized for low electrode losses, at least 1.5 volts of external bias is needed to obtain reasonable electrolytic yield.

Semiconductors may be substituted for the usual metallic electrodes. An n-type semiconductor at the anode and/or p-type at the cathode permits reduction or elimination of the external bias when the semiconductors are illuminated. The band diagrams of a double semiconductor photoelectrolytic cell is shown in Fig. 4.27. The semiconductor electrolyte interfaces are essentially similar to Schottky barriers, as in the liquid junction photovoltaic cell discussed in 4.3.4. Here we are concerned only with short circuit current operation for maximum H_2 generation, however.

The semiconductor anode and cathode may be of different materials but each must be stable against anodic or cathodic corrosion as the case may be. The problem is most severe for the anode, since almost all known n-type semiconductors are unstable against anodic oxidation, the exception being wide-gap oxides such as TiO_2 and $SrTiO_3$ The bandgaps of these materials, at 3.0 and 3.2 eV respectively, are too high to provide a good match to the solar spectrum. The minimum bandgap is set by the requirement

$$E_g \geqslant 1.23eV + V_{BB} + V_{LOSSES} \tag{4.5}$$

where V_{BB} is the band bending voltage used for charge separation in the semiconductor depletion layer. In practice several tenths of an eV for V_{BB} gives adequate current efficiency provided the minority carrier diffusion lengths are greater than the optical absorption depth. Elec-

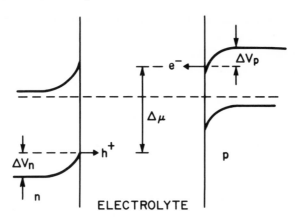

Figure 4.27 Band diagram of a photoelectrolytic cell intended for H_2 production from H_2O, using semiconductor anode and cathode. The available potential for charge separation is ΔV_n (or ΔV_p) in the semiconductor anode (cathode). The maximum free energy available for reduction of water is then $(E_{gn} - \Delta V_n - \Delta E_{Fn}) + (E_{gp} - \Delta V_p - \Delta E_{Fp})$. ΔV_n and ΔV_p are determined by the electron affinities of the semiconductors and chemical potential of electrons in the electrolyte (equivalent to a metal work function) just as for Schottky barriers.

trode overvoltages and other losses suggest $E_g \geqslant 2.2$ eV is needed for a practical cell, assuming the electron affinity of the semiconductor places valence and conduction bands optimally relative to the chemical potentials in the electrolyte.

In the case of TiO_2 the Fermi level in the semiconductor and the chemical potential for the Pt hydrogen electrode are apparently equal, so no band bending occurs. The electron affinity of $SrTiO_3$ is about 0.2 eV less and hence a band bending potential of about 0.2 volts is available to separate photoexcited holes and electrons. The pH of the electrolyte affects the position of the chemical potentials, but the maximum band bending of 0.2 volts for $SrTiO_3/H_2O$ occurs above pH 13. Unfortunately the high bandgap of $SrTiO_3$ restricts the conversion efficiency to less than 0.1 percent for white light, although ~ 10 percent efficiency for narrow band UV light (3200 Å) is observed without external bias [46].

If a p-type semiconductor cathode is used as well the voltages combine in series, so that lower gap materials may be used. GaP is apparently stable under cathodic conditions and provides sufficient voltage to bias TiO_2 to achieve band bending. A maximum power efficiency of 0.25 percent has been measured (ratio of available energy from combustion of the generated hydrogen to total optical energy incident on the cell) for AM2 conditions [47]. N-type GaP cannot be used as the anode because of corrosion. A short diffusion length in the

TiO_2 tends to limit the current efficiency of this cell configuration, together with a higher cathode overvoltage than is obtained with the standard Pt hydrogen electrode.

An additional practical difficulty in cells of this sort is the need to sweep the O_2 generated out of solution. Dissolved O_2 is reduced at the cathode to OH^- in competition with reduction of H_2O to H_2. Oxidation of H_2 at the anode is relatively less important owing to the low solubility of H_2 in water. The physical separation of the oxygen and hydrogen evolved must, of course, be maintained to prevent an explosive hazard. An alternative is to add an ionic species which is oxidized more easily than water. Acetate and butyrate ions in acidic solutions are oxidized to form gaseous hydrocarbons (ethane, etc.) and CO_2. This avoids the explosion problem, the dissolved O_2 problem, and provides an additional fuel output. It has been suggested that the effluent from fermentative biomass conversion is rich in butyric and acetic acid and would provide a cheap input for a photoelectrolytic hydrogen-hydrocarbon plant. [48].

The theoretical maximum efficiency of the direct photoelectrolytic cell is about 23 percent under optimistic assumptions of ideal electrodes. This compares to an efficiency of 16 percent overall for the combination of 27 percent efficient photovoltaic cells with traditional electrolysis (60 percent efficiency) for hydrogen generation, so direct photoelectrolysis is clearly the most efficient fuel-generation pathway *in principle*. The prospects for finding semiconductors with the necessary electrical properties which are stable in water and require only the addition of cheap, readily available electrolyte modifiers do not seem promising, however, and the pace of research in this area has slowed during the past year.

4.4.4 Pyroelectric Generation

The *pyroelectric effect* takes place in dielectric crystals which do not have inversion symmetry as a property of their lattice structure. (Essentially this means that a model of the crystal and its mirror image are distinguishable for some relative orientation of the mirror image and crystal model). This feature permits such crystals to have a permanent electric dipole moment, or spontaneous polarization, $\vec{P}_s$. Ferroelectric crystals undergo a rapid change with temperature of $\vec{P}_s$ and dielectric constant. Thin sheets of such material have been used as infrared detectors, and they could in principle be used as solar power generators. Technically this effect is an example of solar thermal electric generation, but technologically it bears more relation to photovoltaics than to turbine engineering, and we have included it here since it provides, essentially, a solid state electronic device.

A proposed modification [49] of this effect uses a dielectric crystal which is not pyroelectric but which also has a rapid change of dielectric constant with temperature. In this case an external bias is applied to take the place of $\vec{P}_s$. This structure uses LaF_3, an ionic conductor which has an unusually large sensitivity of surface polarization to temperature. The cell consists of a thin slab LaF_3 capacitor with a black absorber coated on one side. Solar radiation is incident on that face and is interrupted periodically; a wobbling mirror could be used to direct the radiation to another cell during the off period. This causes a time varying temperature change in the capacitor, and through the variation in dielectric constant with temperature, a time varying capacitance. At constant charge this results in a time varying voltage.

The equivalent circuit in Fig. 4.28 may be used to model the behavior of either the pyrionic LaF_3 cell or a pyroelectric cell in which the capacitor slab is a ferroelectric such as $LiTaO_3$. For the former material

$$R(T) \propto \frac{1}{C(T)} \propto \exp(U/kT) \tag{4.6}$$

with $U \sim 0.5$ eV, while for $LiTaO_3$

$$R(T) = 0 \tag{4.7}$$

and

$$1/C(T) \propto (T_c - T) \tag{4.8}$$

with T_c the Curie temperature, $\approx 900K$.

With an optical power P_0 incident onto a concentrator of area A_1, an absorber area A, and an absorption coefficient α one has for the average power absorbed

$$P = \alpha \frac{A_1}{A} \frac{P_0}{2} \tag{4.9}$$

The heat balance equation is

$$C_v A l \frac{dT(t)}{dt} = AP(1 + \cos \omega t) - A(\sigma T^4(\tau) - \sigma_0 T_0^4) - AP_L(t) \tag{4.10}$$

with C_v and l the heat capacity and thickness of the LaF_3, and σ and σ_0 twice the product of the Stefan-Boltzman constant and the cell or chamber wall emissivities. For the circuit equation one has

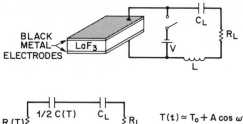

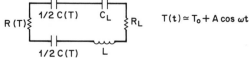

Figure 4.28 Schematic and equivalent circuit of a pyrionic solar electric generator using LaF$_3$ (after Ref. 49, this Chapter).

$$L \frac{dI(t)}{dt} + (R(t)+R_L)I(t) + \frac{q(t)}{C(t)} + \frac{q_L(t)}{C_L} = 0$$

$$(4.11)$$

Solution of 4.2 and 4.3 is possible if one assumes that the average value of the cell temperature T_0 is large compared to the time varying part $\Delta(t) = \Delta_0 \sin \omega t$, and only first order terms are retained. The second law of thermodynamics requires the total entropy change over a cycle for a heat engine such as this to be positive and hence limits the efficiency. This thermodynamic restriction gives as an upper limit for the efficiency η

$$\eta_+ < 1 - \frac{T_0}{T\sqrt{1-(\Delta_0/T)^2}}$$

$$(4.12)$$

For $T_0 = 300K$, $T = 500K$ and $\Delta_0/T \sim 0.1$ this gives 40 percent. Note that the limiting efficiency increases with operating temperature of the absorber as is typical for heat engines.

The other limit on efficiency is set by the specific principles of the device operation. A lower limiting efficiency can be estimated [49]:

$$\eta_- \sim \frac{1}{2} \left[1 + \frac{1}{4\zeta} - \sqrt{(1 + \frac{1}{4\zeta})^2 - 1} \right]$$

$$(4.13)$$

with

$$\zeta = \frac{P}{2R_0A} \left(\frac{VU}{2C_v lkT^2\omega} \right)^2$$

$$(4.14)$$

where V is the bias voltage to which the cell was charged initially, R_0 is the time average value of $R(\tau)$, and it has been assumed that R_L was chosen to match R_0 and L chosen such that $LC_L = \omega^{-2}$, i.e., the circuit was resonant. For $\omega = 30 \sec^{-1}$, $A_l/A = 54$, $l = 10 \ \mu m$, $V = 3$ volts and other values as measured for LaF$_3$ [50], $\zeta = 1$ and an efficiency of 25 percent is predicted. A rather large inductance would have to be used for resonance, suggesting an active phase shift network rather than an inductance coil, but this need not pose a major problem. The power is delivered at ~ 5 Hz ac so power conditioning to dc or to 60 Hz ac is required. No working models have yet been constructed and the material aspects of LaF$_3$ production and device fabrication have not yet been addressed. Nevertheless the anticipated efficiencies are similar to those expected for single crystal concentrator photovoltaic cells and the power is available at higher voltage ($\sim 52V$ peak) and lower current than with solar cells. Cooling of the low temperature bath (chamber walls) is required, but this should pose less of a problem than cooling the small area concentrator cells. The possibility that superior ionic conductors to LaF$_3$ may be found cannot be discounted.

The pyroelectric version using a material such as LiTaO$_3$ has not been analyzed in detail. One anticipated difficulty would be the tendency for the ferroelectric to depole; i.e., to break $\vec{P}_s$ into domains of opposite sign. If operated *above* the Curie point a large change of dielectric constant with temperature can also be obtained, but it is not known whether sufficiently blocking electrodes and high surface polarization would also be available. In the centrosymmetric high temperature phase an external bias charge would have to be supplied occasionally as with LaF$_3$.

4.5 Summary

To predict the timing or even occurrence of success in a research project is almost by definition impossible. Each of the variety of approaches discussed in this chapter has at least one compellingly attractive aspect, and each presents at least one unsolved problem of a potentially fundamental and perhaps insurmountable character. Many of these problem areas coincide with developing areas in fundamental solid state physics research and continued research effort would be desirable apart from the motivation provided by the need for solar energy development. Work to elucidate the nature of surfaces and interfaces of semiconductors, the nature of electronic states in amorphous materials, the properties of ionic conductors and the limits on conductivity in organic films, for example, has engaged the attention of some of the best scientific minds in the world today. Perhaps the solar electric technology of the future will evolve from basic research in these areas along very different lines from traditional photovoltaics. The proposed pyronic cell, for instance, provides a good example of a truly innovative, out-of-the-main-stream concept which cannot be dismissed out of hand as impossible.

One must realize that even if scientific feasibility can be shown for any of the approaches in this chapter (which we may take as achievement of a laboratory cell efficiency in the 12 percent range), very serious development and engineering problems remain. The non-ingot silicon efforts depart least from the well-known (some would say well-beaten) track, and hence probably offer the best prospect of success, although the success will be of an incremental nature rather than a major breakthrough. On the other hand realization of a practical photovoltaic cell based on an organic thin film material may be regarded as most problematic. This would necessitate a major upheaval in our present knowledge of electronic states in organic compounds. Direct hydrogen generation appears less feasible than electric power generation with photoelectrolytic cells, but the pressing need for energy in portable fuel form will continue to provide motivation for research in that direction. Liquid-junction cells do not now appear to be competitive simply as photovoltaic generators.

Finally one must realize that we are not involved in a solar electric race that will be over once and for all when DOE goals are met, with whatever technology that may be. There will *always* be a need for improved performance where energy generation is concerned, which will only be realized through appropriate support of basic research of both the evolutionary and revolutionary kind.

REFERENCES

1. T. F. Ciszek and G. H. Schwuttke, *Phys. Stat. Sol. (a)* **27**, 231 (1975).

2. R. G. Seidensticker, *J. Cryst. Gr.* **39**, 17 (1977).

3. J. D. Zook, S. B. Schuldt, R. B. Maciolek and J. D. Heaps, *Proc. 13 PVSC,* 472 (IEEE, New York, NY, 1978).

4. K. V. Ravi, *J. Cryst. Gr.* **39**, 1 (1977).

5. I. A. Lesk, A. Baghdadi, R. W. Gurtler, R. J. Ellis, J. A. Wise and M. G. Coleman, *Proc. 13 PVSC,* 173 (IEEE, New York, NY, 1978).

6. D. L. Barrett, E. H. Mayers, D. R. Hamilton and A. I. Bennett, *J. Electrochem Soc.* **118**, 952 (1971).

7. J. D. Heaps, R. B. Maciolek, J. D. Zook and M. W. Scott, *Proc. 12 PVSC,* 147 (IEEE, New York, NY, 1978).

8. T. L. Chu, S. S. Chu, Roshdy Abdrresool, C. L. Lin and E. D. Stokes, *Proc. 13 PVSC*, 1106 (IEEE, New York, NY, 1978).

9. T. L. Chu, *J. Cryst. Gr.* **39**, 45 (1977).

10. P. H. Robinson, D. Richman, R. V. D'Aiello and B. W. Faughman, *Proc. 13 PVSC*, 1111 (IEEE, New York, NY, 1978).

11. M. W. Geis, D. C. Flanders, and H. I. Smith, *Appl. Phys. Lett.* **35**, 71 (1979).

12. G. D. Boyd, L. A. Coldren, and F. G. Storz, Paper N-7, 21st Electronic Materials Conference, Boulder, CO. (June 1979, unpubl.).

13. D. Carlson, C. R. Wronski, A. R. Triano and R. E. Daniel, *Proc 12 PVSC*, 607 (IEEE New York, NY, 1977).

14. N. Mott, Science **201**, 871 (1978).

15. R. Fisch and D. C. Licciardello, *Phys. Rev. Lett.* **41**, 889 (1978).

16. S. R. Ovshinsky and A. Madan, *Nature* **276**, 482 (1978).

17. C. R. Wronski, *Proc. 13 PVSC*, 744 (IEEE New York, NY, 1978).

18. D. E. Carlson and C. R. Wronski, *Appl. Phys. Lett.* **28**, 671 (1976).

19. R. Vohl, D. M. Perkins, S. G. Ellis, R. R. Addis, W. Hui and G. Noel, *IEEE Trans. ED* **14**, 26 (1966).

20. L. M. Fraas and K. Zanio, *Hughes Research Report #521*, (1978, unpubl).

21. W. D. Johnston, Jr. in *Conversion and Storage of Solar Energy*, R. B. King, C. Kutal and R. R. Hautala, Eds., p. 237 (Humana Press, Clifton, NJ, 1979).

22. J. J. J. Yang, P. D. Dapkus, A. G. Campbell, R. D. Yingling and R. D. DuPuis. Paper A-3, 20th Electronic Materials Conference, Santa Barbara, CA (June, 1978, unpubl.).

23. cf. Ref. 20, 21, 24 and 25.

24. Y. C. M. Yeh and R. J. Stirn, *Appl. Phys. Lett.* **33**, 401 (1978).

25. S. S. Chu, T. L. Chu and H. T. Yang, *Proc. 13 PVSC*, 956 (IEEE, New York, NY, 1978).

26. R. J. Stirn, Y. C. M. Yeh, E. Y. Wang, F. P. Ernest, and C. J. Wu, *1977 IEEE Int. Electron Device Meeting Tech. Digest*, p. 48 (IEEE, New York, NY, 1977).

27. A. E. Blakeslee and S. M. Vernon, *IBM J. Res. Dev.* **22**, 346 (1978).

28. S. Chu (April 1979, unpubl.).

29. Y. C. M. Yeh, F. P. Ernest and R. J. Stirn, *Proc. 13 PVSC*, 966 (IEEE, New York, NY, 1978).

30. W. D. Johnston, Jr. and W. M. Callahan, *J. Electrochem Soc.* **125**, 977 (1978).

31. J. P. Duchemin, M. Bonnet, G. Beuchet and F. Koelsh, Proc. 1978 Conf. on GaAs and Related Compounds (Inst. of Phys. Conf Series **45**, Inst. of Phys, London, UK, 1979).

32. A. A. Barybin, V. A. Kempel' and V. V. Nechaev, *Izv. Akad. Nauk SSSR, Neorg. Mat.* **13**, 42 (1977).

33. J. L. Shay, M. Bettini, S. Wagner, K. J. Bachmann and E. Buehler, *Proc. 12 PVSC,* 540 (IEEE New York, NY, 1977).

34. K. Sreeharsa, K. J. Backhmann, P. H. Schmidt, E. G. Spencer and F. A. Thiel, *Appl. Phys. Lett.* **30,** 645 (1977).

35. K. C. Chang, A. Heller, B. Schwartz, S. Menezes and B. Miller, *Science* **196,** 1097 (1977).

36. A. Heller, B. A. Parkinson and B. Miller *Proc. 13 PVSC,* 1253 (IEEE New York, NY, 1978).

37. W. D. Johnston, Jr., H. J. Leamy, B. A. Parkinson, A. Heller, and B. Miller, *J. Electrochem. Soc.* **127,** 90 (1980).

38. A. Heller, B. Miller, S. S. Chu, and Y. T. Lee *(J. Amer. Chem. Soc.,* in press).

39. L. L. Kazmerski, P. J. Ireland, F. R. White and R. B. Cooper, *Proc. 13 PVSC,* 184 (IEEE New York, NY,, 1978).

40. D. Trivich, E. Y. Wang, R. J. Komp and Anand S. Kakar, *Proc. 13 PVSC,* 174 (IEEE New York, NY, 1978).

41. A. Catalano, V. Dalal, W. E. Devaney, E. A. Fagen, R. B. Hall, J. V. Masi, J. D. Mealsin, G. Warfield, N. Convers Wyeth and A. M. Barnett, *Proc 13 PVSC,* 288 (IEEE New York, NY, 1978).

42. Y. Y. Ma, A. L. Fahrenbruch and R. H. Bube, *Appl Phys. Lett.* **30,** 423 (1977).

43. N. Nakayama, H. Matsumoto, K. Yamaguchi, S. Ikegami and Y. Hioki, Japan, *J. Appl. Phys.* **15,** 2281 (1976).

44. V. Y. Merritt and H. J. Hovel, *Appl. Phys. Lett.* **29,** 414 (1976).

45. H. Gerischer, *Electroanal. Chem. and Interfacial Electrochem.* **58,** 263 (1975); A. Fujishima and K. Honda, *Nature* **238,** 37 (1972).

46. J. G. Mavroides, J. A. Kafalas and D. F. Kolesar, *Appl. Phys. Lett.* **28,** 241 (1976).

47. A. J. Nozik, *Appl. Phys. Lett.* **29,** 150 (1976).

48. R. F. Schwerzel, E. W. Brooman, R. A. Craig, D. D. Levy, F. R. Moore, L. E. Vaaler and V. E. Wood in *Conversion and Storage of Solar Energy,* R. B. King, C. Kutal and R. R. Hautala, Eds, p. 83 (Humana Press, Clifton NJ, 1979).

49. A. Sher, *Proc. 12 PVSC,* 1000 (IEEE New York, NY, 1977).

50. A. Sher, C. L. Fales and J. F. Stubblefield, *Appl. Phys. Lett.* **28,** 676 (1976).

CHAPTER 5

Beyond the Cell

Ultimately, the market for photovoltaics is not a market for solar cells but a market for solar electric *systems*. The consumer, be it a private individual, an industrial corporation, or a public electric utility, will wish to purchase electric power when and where it is needed with the assurance of uninterrupted availability. The questions of reliability and availability of photovoltaic power play a major role in determining the ultimate practicality of this form of solar energy. The same year the first practical solar cell was announced (1954), initial field tests were carried out to establish the weatherability and practical current generating capability of silicon solar cell arrays as a potential source of terrestrial electric power [1]. An initial finding was that eight days storage is needed to assure that power to a steady load at 10 percent of the maximum (bright day, noon) output can be met. This is consistent with the findings of numerous subsequent studies of a much more extensive and sophisticated nature. This early study also served to confirm the intrinsic reliability of the silicon cell elements, and pointed up the need for a weatherproof encapsulated assembly to protect the contacts and interconnections between cells.

The assembly of interconnected and encapsulated cells is called a *module*. As the name implies, one should be able to combine modules readily to form an *array* of desired size or capacity. The array, together with appropriate impedance matching (power conditioning) and storage equipment, constitutes a photovoltaic system. In this chapter, we will consider the requirements which are imposed on module designs and the approaches being explored today to meet those requirements within the DOE cost goals and still assure an expected operational life exceeding 20 years.

Concentrator modules which contain focusing optical elements or other components to increase the useful radiation incident on the solar cells offer substantial design flexibility. These may be of an imaging or nonimaging type. This approach permits significant reduction in semiconductor area and hence allows high performance, expensive cells (such as single crystal GaAs-based heterojunction cells) to be considered, but at the same time mechanical and optical complexity is introduced, adversely affecting reliability.

Finally, the needs and possibilities for power conditioning and storage are discussed in this chapter as the modules and arrays must obviously be designed for compatibility with the power-accepting equipment.

5.1 Flat-Plate Modules

The flat-plate solar cell array is conceptually the simplest arrangement for photovoltaic power generation. It is an arrangement compatible with roof top installation. Flat-plate arrays can be designed to perform the functions of ordinary roofing materials [2], which reduces the effective net cost of the installation. Array design is primarily a matter of structural and mechanical engineering. Strong, waterproof, lightweight and fire-resistant assemblies are needed, with sufficient ruggedness to permit installation by nontechnical craftspeople in the construction trades.

The optimum orientation for arrays of this type is south-facing, at an elevation angle equal to the site latitude give or take an adjustment depending on seasonal load variations. An elevation increase to site latitude *plus* winter solar declination will provide a more constant power output throughout the year, for example, at the cost of maximum summer power generation. Angles of 40° to 60° are thus to be expected in the United States; this corresponds to a *steep* roof which will tend to be self-cleaning of snow or other forms of precipitated deposits. A smooth, hard array top surface (such as glass) encourages this tendency. As an alternative to fixed installation, a structure permitting seasonal adjustment of the inclination angle may be considered. This adds little to the cost of free-standing or roof-mounted arrays but is probably not practical for an array intended to replace conventional roofing materials. Fully tracking (two-axis) flat plate arrays are generally considered to be impractical. The gain in power generation from tracking is partly offset by the larger area required to prevent shadowing. The improvement over a fixed flat plate array oriented optimally for the location and season is less than a factor of two, not enough to offset the tracking costs. The flat plate approach emphasizes simplicity and adequate performance at low cost, most of which goes to cell area. Concentrator modules on the other hand *require* two-axis tracking for economic viability and place most of the cost in the mount and optical elements rather than in the cells.

Another factor to be considered in module design is the need for heat transfer and rejection. Free-standing arrays permit convective heat removal from a finned back surface. Arrays formed as part of roof structures do not provide for an easy solution in this respect. For free standing Si flat plate arrays, operating temperatures of 40 to 60C have been determined for 25C ambients and passive convective cooling in still air. This is not enough to cause a *large* reduction in efficiency, but the smaller temperature rise of 10 to 20C observed with 5 to 10 mph breezes is to be preferred. Evidently standards, or ratings, of photovoltaic modules should be developed in a fashion that permits evaluation of performance under actual installed conditions rather than the standard conditions used for reporting cell efficiencies.

5.1.1 Materials for Flat-Plate Modules

Since a 10 percent efficient flat-plate module will produce 100 W_{pe}/m^2 (~10 W_{pe}/ft^2) a cost total of $50/m^2$ ($5/ft^2$) is required to meet the interim $.50/W_{pe}$ goal. Since the photovoltaic cells will represent a large part of this cost, the materials for module assembly will have to be very cheap indeed, totaling perhaps $10/m^2$ or less. At minimum a front and a back layer are required together with cell interconnect metallization and power output terminals. Either the module front or back or both must be *structural*, i.e., provide adequate rigidity to permit edge joining of the module to other modules in the array frame.

Representative costs of some candidate materials are listed in Table 5.1. It is apparent that design of a suitably low cost encapsulating sandwich for photovoltaic modules is a nontrivial task, particularly since glass and steel sheet are heavily mass produced and future cost reduction is improbable. The present commercial modules (see Fig. 5.1) use either aluminum sheet or fiberglass-epoxy composite backing with either RTV silicone or boro-silicate (low-iron) glass front covering. Without allowing for assembly labor the materials costs alone are in the

Table 5.1
Cost of Potential Encapsulating Materials — $/ft^2

Front Surface			
Structural (sheet)		Nonstructural (films)	
glass	0.25—0.75	Acrylic	0.01—0.04
acrylic	1.75—2.00	Silicone	0.04—0.06
		Fluorocarbons	0.10—0.20
		AR coating	0.04

Potting		Adhesives	
RTV	3.00	acrylic	0.04
silicone gel	1.00	epoxy	0.03
PVC	0.03—0.10	silicone	0.10
polyvinyl butyral	0.20—0.35		
urethane	0.07—0.10		
acrylic	0.10—0.15		

Back Surface			
Structural (Sheet)		Nonstructural (film)	
glass	0.25—0.30	Tedlar	0.10
aluminum	1.00 +	Mylar	0.10
stainless steel	3.00 +	aluminum	0.04
enameled steel	0.50—0.75	stainless steel	0.04
plywood	0.14		
presswood	0.14		
honeycomb paperboard	0.10		

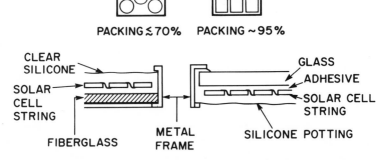

Figure 5.1 Contemporary photovoltaic module design. (a) High areal efficiency requires rectangular cells. (b) Glass cover sheets have proved more resistant to atmospheric pollutants and soot than modules faced with silicone rubbers.

$2 to $4/ft^2 range. These might possibly be compatible with the near-term $2/W$_{pe}$ cost goal, but present emphasis has shifted toward glass sheet for the front and either enameled steel or soda-lime glass sheet for the backing.

Contemporary modules include a potting compound to cushion the interconnected cells between the front and back covers, the edges of which are sealed with an elastomer gasket into a steel channel frame. It is possible, however, to bond glass sheets to silicon, to aluminum foil interconnects, and to each other without the use of organic adhesives [3]. This is accomplished by field-assisted bonding (FAB) which provides an electrostatic bond exceeding the yield strength of either Si or glass. The mechanism of bond formation is shown in Fig. 5.2. At elevated temperatures $\geq$ 400C the alkali ions (Na^+, Li^+) in the glass become mobile and are drawn away from the bond interface. Excess O^- ions left behind interact to form an oxide compound with the semiconductor or (oxide-forming) metal. A temperature in the neighborhood of 600C permits the glass to deform readily around the cells and metal foil conductors. When the module is cooled, an all glass hermetic seal results, without the need for any organic potting material or adhesive to add additional degradation modes. Output terminals may be fabricated by aluminum foil strips bonded on both sides to the glass,

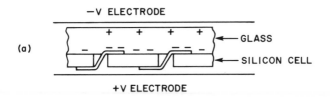

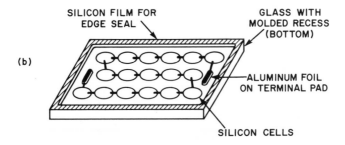

Figure 5.2 (a) Schematic of field-assisted bonding of glass to silicon. (b) View of lower half of all-glass module encapsulation. In practice the cells and interconnect foils would be bonded first to the flat top glass plate (not shown).

for example. Contact to these may be made through preformed holes in the back cover, allowing transition to a non-oxide forming metal terminal. The glass-to-glass edge bonds require that a thin conducting layer be predeposited on the glass at the bonding surface; this is readily accomplished with an aluminum evaporation.

A study of the economics of such an all-glass module system concluded that a manufacturing cost of $1/ft^2 covering the materials, labor and capital equipment amortization was plausible [4]. The primary uncertainty is the price of suitable boro-silicate front-surface glass which is now much too expensive, and the $1.00/ft^2 result depends on a cost of $0.30/ft^2 being reached in mass production. (Soda lime glass [window glass] costs ~$0.25/ft^2 at wholesale today). The requirements on the FAB glass, beyond appropriate optical performance required of any glass front cover design, are that the thermal expansion coefficient match silicon closely and that the yield point for adequate plastic flow be low enough that the silicon cells are not heated excessively. Corning 7070 type glass has been found to be suitable [5].

CdS/Cu$_2$S flat plate modules pose different encapsulation problems from those presented by Si wafer cells, but the goal is the same: complete protection from humidity and chemical contamination for at least 20 years in an outdoor environment. Present commercial construction [6] combines module formation with cell fabrication so that only the bonding of a tempered glass front cover to the epoxy covered steel back sheet is required. These modules are typically 20×20 cm (8×8 inches) and are assembled into panels on extruded aluminum frames. The frames provide weather protection for soldered series/parallel connections between modules. Array fabrication costs for CdS/Cu$_2$S cells appear to be much less an issue than achievement of the 10 percent array efficiency goal.

5.1.2 Electrical Assembly of Modules

The interconnection of silicon wafer solar cells within a module is in principle a simple problem, but in practice is rather more complicated than it might at first appear. A series connection eases the problem of handling high currents but requires a higher quality of insulation between the cells and the metallic frame or backing members. In a series string the cells must be well matched for current output at uniform illumination, and the array siting must be such as to minimize the chance of partial shading of the string. When one cell is not illuminated or for other reasons does not generate the same photocurrent as others in a series string, the output of the entire string is sharply reduced. A nonilluminated cell is reverse biased and will undergo destructive breakdown if the string voltage exceeds the avalanche breakdown voltage of the junction. Appropriate protective circuitry can be incorporated as a safety measure, of course, but at added cost and complexity.

An arrangement in which the cells within a module are connected in parallel makes the matching much easier, since the cell operating voltages vary only logarithmically with intensity variation and cells of the same design typically show much less variation in voltage output than current output characteristics. However, one would like to have as a practical minimum output for a module a voltage compatible with battery charging, typically 14 to 15 volts for use with automotive-type lead acid batteries. For that reason most commercially available Si terrestrial modules have consisted of 30 to 40 series connected cells. The modules are then connected in parallel to form panels of up to 20 to 30 ampere capacity. These panels are finally series connected into arrays providing several kW of peak power.

From the point of view of manufacturing cost it would be preferable to have a significant number of cells connected in parallel as elements of the first series strings, so that averaging over the individual cell current outputs would reduce matching problems. Insulation and breakdown problems are also reduced, but a primarily series connection of modules to form panels is then required. This requires high-current, low impedance inter-module connectors.

The actual connection of cells to each other and to a parallel current bus is usually accomplished by welding or soldering a silver, aluminum, or copper foil to the metallized cell contact areas. This can be performed by automated machinery but care must obviously be taken to ensure that the solar cell wafers are not cracked or damaged by excessive heat or pressure at the contact joining. Metallization deterioration at bonding pads is a notorious cause of semiconductor device failure and the cell interconnection interface appears at present to be the weak link overall in module reliability. Redundant interconnection is provided by several contemporary module manufacturers as the simplest solution to this problem.

5.1.3 Weatherability and Soiling of Flat-Plate Arrays

The first field test [1] of a silicon solar cell array occurred near a nesting site of large birds, and provided initial experience with one form of array soiling often suggested as a potential difficulty for photovoltaic power systems. However, the fact that arrays are exposed to the elements implies they will be washed naturally at more or less regular intervals by rainfall. Practical experience shows this is adequate to remove soiling from most natural sources. Recent tests, however, have shown that there is a significant difference between modules with glass front encapsulation and those made with a transparent silicone potting compound as regards recovery from soiling.

As shown in Fig. 5.3, glass covered modules lose about 5 percent of the original output within a few days but remain at that level, while the other modules suffer an initially larger loss which continues to increase with time. The glass covered modules recover to the as-installed performance level after washing, while the silicone material does not fully recover and hence these modules suffer a permanent loss in performance. A recent study [7] of modules exposed to the environment in seven test sites including tropical rain forest, desert, sea air and urban industrial environments showed that, while soiling effects were dramatically worse for the silicone covered cells exposed to industrial air pollutants, the glass covered modules behaved similarly in all environments and always suffered less performance loss than did the silicone covered cells. The silicone material was found to be prone to encrustation with mossy plant growth or algae in the rain forest environment. In this study some delamination of the silicone covered modules was observed, and several of the glass cover sheets cracked during handling. Neither of these effects produced any apparent output loss, but it is reasonable to assume that over a longer period of time corrosion and failure would ensue.

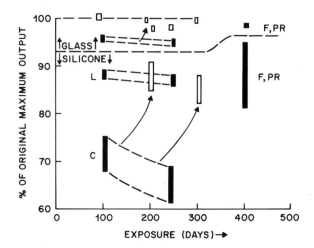

Figure 5.3   Loss in output due to soiling for glass and silicone protected silicon solar modules. Site locations: C — Cleveland (industrial area), L — Cleveland (suburban area); F,PR — Florida and Puerto Rico. Open bars indicate output after post-exposure washing (data from Ref. 7, this Chapter).

Another recent study [8] addressed the effects of humidity, thermal cycling and ultra-violet exposure on silicon wafer cells encapsulated with acrylic and silicone front coatings, polymeric film and sheet front covers (polyester or acrylic sheet, etc.), or three types of adhesively-bonded glass front covers. *All* forms of encapsulation failed to provide protection against exposure to high humidity (97 percent RH at 38C). In the case of the glass encapsulated seals this appeared to be due to failure at the seal edges which incorporated silicone adhesives. The deterioration was manifest predominantly as an increase in series resistance and an associated drop in fill factor. The cells in this study were fabricated with silk-screened Ag contacts, which are known to be more sensitive to humidity than other contact configurations (e.g., Ti/Pd/Ag).

Overall, it appears that glass front covers are required for soiling resistance, and a full, hermetic seal to the back cover, be it another sheet of glass, or enameled steel, appears to be required for resistance to humidity. Silicone adhesives and potting compounds do not provide the performance which is thought necessary to assure a 20-year life under normal operating conditions. Alternative sealing measures such as electrostatic bonding or glass frit sealing will have to be developed. On the positive side, it appears that no regular maintenance for glass-fronted modules is required except possibly in locations where snow and ice or unusually heavy avian activity are encountered.

5.1.4 Panel and Array Support Structures

Flat panel solar arrays installed today typically consist of frame assemblies of modules mounted on simple truss or beam supports (Fig. 5.4). This approach is simple, convenient, and provides ready access to the modules for maintenance, but it has cosmetic disadvantages in residential applications. The cost of truss or beam mounting increases rapidly with the height of the supported panels above the ground (or flat roof), and with panel weight and wind loading allowances. This type of mounting seems appropriate above parking lots or on the (flat) roof-tops of light industrial buildings or shopping centers, and would not require acquisition of additional land in such instances.

Residential installation poses other problems. Direct mounting on existing, adequately sloped and oriented roofs or integration into new building structures can be used to save material, labor and land costs. The last consideration is becoming nontrivial in many parts of the United States. Suburban building lots in many east coast communities

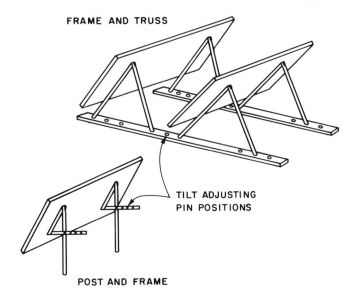

FRAME AND TRUSS

TILT ADJUSTING
PIN POSITIONS

POST AND FRAME

Figure 5.4   Truss and post frame mounting configurations. The truss allows for simple seasonal adjustment of array tilt by coupling the support braces to a second, movable stringer (not shown), and is the cheapest frame support system.

sell at more than $45,000/acre ($1.00/ft^2), and in certain prestigious communities in California prices are as much as ten times higher. As an *approximate* size estimate, 100 m^2 (1000 ft^2) of array will produce 1200 kWhrs/mo. on average. Roughly 75 m^2 (750 ft^2) will be required of horizontal base surface lying directly beneath the cells. If a truss mounting is used about 2500 ft^2 of horizontal surface is required to prevent shading — a significant portion of a fractional acre lot.

For new construction architectural designs can be employed which permit direct enclosure of roof or sloped wall elements with photovoltaic modules. Several possibilities are shown in Fig. 5.5. Exploratory development [9] has led to fabrication of samples of solar cell shingles for evaluation as a roofing element in new construction (Fig. 5.6).

Little experience has yet been gained with such integrated construction, largely because present day photovoltaic modules are too expensive for use in actual residences, and have not been designed for wall or roof integration. An important concern will be the cell temperature rise when modules are installed on thermally insulated roof or wall structures. Vandalism, lightning, and damage from inclement weather are risks that will have to be underwritten. Demonstration programs subsidized by the U.S. Department of Energy are expected to begin providing practical information on the extent of these problems in the residential area soon.

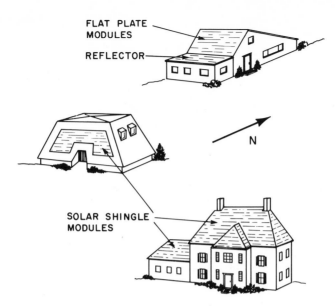

Figure 5.5 Architectural possibilities for contemporary, modern and traditional residential construction using photovoltaic modules in roof or wall elements.

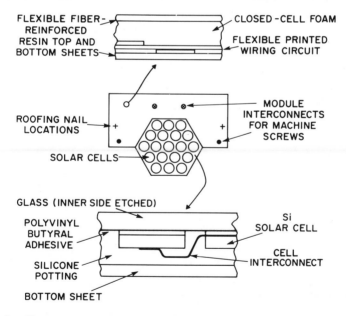

Figure 5.6 Hexagonally packed roofing-shingle solar cell module (after Ref. 9, this Chapter). Some parameters of interest are: specific output 98 W/m^2 when installed on an insulated roof surface at 0.8 kW/m^2 insolation, 1 m/sec wind velocity and 20C ambient. Shingle weight is 4 lbs/ft^2 (19.7 kg/m^2) including overlap, comparable to top-grade asphalt shingles.

5.2 Concentrator Modules

A primary source of difficulty in attaining efficient conversion of solar energy is the dispersed nature of sunlight as an energy source. Solar cells can be designed to operate at intensities of several hundred to several thousand times that provided by unfocused solar radiation. For properly designed cells efficiencies somewhat higher than the best that can be obtained without concentration have been achieved. The available electric power per unit area of solar cell surface is thus increased by a factor equalling or even slightly exceeding the concentration or focusing ratio. This can be true even after allowance is made for moderate losses in the concentration system. Evidently this is a cost-effective approach if solar concentration systems can be built less expensively than flat plate photovoltaic modules of the same aperture area. Expensive, high performance solar cells can be used at high concentration ratios while the total system cost is still dominated by the light gathering elements.

There are two main types of solar concentrators. *Direct* concentrator systems include arrangement of lenses and/or mirrors to focus the sunlight passing through the system aperture onto the solar cell. For these the concentration ratio is the ratio of the entrance area to the exit area of the optical system, multiplied by the optical transmission factor. *Indirect* concentration systems may or may not include lens or mirror arrangements. They differ from direct systems in that the solar radiation is used to excite a secondary radiating medium. Re-emitted radiation from the secondary radiator is then converted to electricity in photovoltaic cells. The advantage of the two-step process is that the secondary radiator may be chosen to provide a different spectral distribution from that of the incident sunlight; peaking, for instance, at the wavelength most efficiently converted by the photovoltaic devices used.

5.2.1 Principles of Direct (Optical) Concentrators

Direct optical concentrators accept sunlight entering as a well collimated bundle of rays through a large aperture and condense this to pass through a small exit aperture. In the process the beam must become divergent, since *brightness* cannot be increased in a refractive or reflective optical system. Brightness is a property of the radiation source, the sun in this case, and is measured in watts per cm^2 per steradian of divergence. An optical concentrator must be pointed exactly at the sun for *all* the rays collected to pass through the minimum sized exit aperture. The angular tolerance varies inversely with the concentration ratio. For imaging optical concentrators, *no* output is obtained if the tracking error exceeds the angle subtended by the sun $(4.7 \times 10^{-3}$ radians), since this error moves the solar image all the way off its design position.

There are nonimaging forms of optical concentrators as well [10]. These cannot reach the theoretical maximum ratios of imaging systems, but do allow the possibility of attaining a more uniform intensity distribution across the exit aperture and an improved tolerance to tracking errors. It can be shown that the relation between maximum concentration and acceptance half-angle θ_c is just

$$C = \frac{1}{\sin^2\theta_c} \qquad (5.1)$$

for point concentrators, or

$$C = \frac{1}{\sin\theta_c} \qquad (5.2)$$

for line concentrators, given ideal optical elements. Real optical elements are not ideal. Mirror surfaces deviate from the design curve, and lenses have inherent aberrations as well as manufacturing imperfections. For photovoltaic concentrator modules the aperture cost must be held below a few dollars per ft^2 and hence high precision optics are not feasible.

The performance of photovoltaic concentrator cells falls rapidly if a given radiation flux is distributed nonuniformly. As a practical matter, one would want modules to be of a reasonable size, 1 to 10 ft^2 in aperture, for instance. The cell elements should not be too large either, since inconveniently large currents would then have to be handled by the frame bus. Cell diameters in the range of 1 to 10 cm seem reasonable. One would then want a concentrator to achieve ratios of several hundred while allowing a few degrees spread of angular acceptance and providing an illumination level constant to within a factor of two or better over the cell.

A simple plastic Fresnel lens would seem to be a good solution, but imaging Fresnel lenses are not a good choice because they give a strongly nonuniform photon flux at the focus. This can be improved somewhat by placing the cell *away* from the focus, but a better solution is to design the facetting of the Fresnel sheet to achieve a uniform quasi-focal distribution [11]. This can be done by cutting the various facets, not to reduce spherical aberration, but to achieve an aspheric lens effect (see Fig. 5.7). The spectral response of the cell and the chromatic aberrations of the lens must be considered as well, since uniform generation of photo-current is required. Doming the lens allows

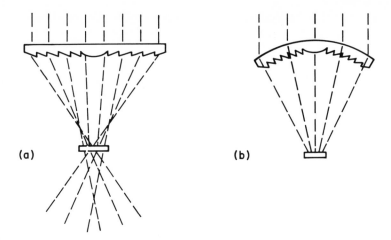

Figure 5.7 Improved Fresnel lens designs: (a) aspheric lens. The outer rays are focused to the center of the target while the middle rays are focused to the edges of the target. (b) Domed lens for high concentration. Aberrations are reduced by utilizing refraction at both surfaces, but molding, cleaning, and corner utilization are problems (cf. Ref. 21, this Chapter).

some refraction to take place at the first surface, but the facet design must be compromised to permit removal from the mold. The design of acrylic Fresnel lenses providing concentration ratios over 10^3 with less than $\pm 30\%$ variation in the uniformity of receiver illumination appears possible. The cost of acrylic sheet is presently $\sim\$2.00/\text{ft}^2$ (see Table 5.1). It is not clear that molded lenses will be cheap enough to permit their use in practical concentrator modules, unless the cost of other components and the concentrator cells can be held to a few dollars per ft^2 of aperture.

Reflective optics may offer advantages in fabrication, particularly in two dimensional, line-focusing concentrators. A simple parabolic trough is an imaging collector, and as such requires good tracking accuracy. The Winston compound parabolic collector [12] (CPC) is a nonimaging variant which is ideal in the sense that all light entering the aperture within the acceptance angle contacts the receiver (Fig. 5.8). It consists of two parabolic surfaces, in the line-focusing case. A related point-focusing concentrator may be generated as in Fig. 5.8 (b) or (c), but neither the combination of two orthogonal CPCs nor the conical surface of revolution act as ideal concentrators [13]. Additionally, these do not provide uniform intensity at the target.

Cassegrain and other multireflector designs can provide good uniformity of illumination and high concentration ratios, but add expense and complexity. Naturally, they are subject to the same limitations as apply

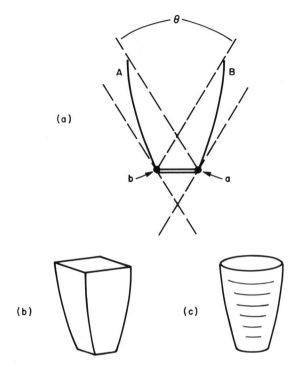

Figure 5.8   (a) Cross-section of Winston compound parabolic reflective trough concentrator. Parabolic surfaces A and B are positioned each with its focus on the other's surface (A' and B'). The acceptance angle θ is the same as the angle between the axes of the parabolas. (b) and (c) Point-concentrating CPCs formed by (b) intersection of two orthogonal CPC troughs, and (c) by rotation of a CPC cross-section to form a conical CPC (see Refs. 11 and 12, this Chapter).

to the other optical concentrator designs and at high concentration ratios require good tracking accuracy. As a practical matter concentrator modules will be mounted on a frame which can be pointed as a whole toward the sun by motors controlled by a sun-position sensor. The sensor may easily be kept pointed toward the sun to within 10^{-3} rad. Most of the practical tracking error is actually due to flexing of the frame by wind, residual error in the alignment of modules on the frame to the sensor axis, etc. These errors, naturally, are related to the cost of the tracking mount and module assembly.

5.2.2 Indirect Concentrators

Two types of concentrators have been proposed in which the photovoltaic cells are not illuminated by sunlight. In the first a flat plate containing a fluorescent material is employed [14] . The second uses a high-ratio solar concentrator to heat a radiator to incandescence, converting the solar (5800K) spectrum to that of a black body at lower temperature [15]. Research-scale experimenting with prototypes of each is now underway.

The luminescent collector (Fig. 5.9) works by absorbing photons from the solar spectrum and emitting longer wavelength photons in randomized directions. By proper design of the luminescent plate, most of these can be trapped inside the plate by total internal reflection. Three of the four edges of the plate are polished and silvered; solar cells are optically contacted to the fourth edge. This concentrator can be operated like a flat plate array, but can theoretically achieve concentrations of several hundred. The best that has been achieved to date in preliminary experiments is about unity, however, because of reabsorption losses in the luminescent medium [16]. The possibility to use inorganic ions in a glass matrix exists, but organic dye molecules in a liquid or plastic medium are more attractive from a cost point of view. One must keep in mind that known organic dyes with high fluorescent yield decompose with a probability of 10^{-3} to 10^{-4} or greater per absorption-emission cycle, so that the compounds used in dye lasers, for instance, are probably not practical for solar luminescent concentrators. One can imagine sheet-like plastic tanks from which the dye solution could be pumped and renewed periodically, but this is certainly much less satisfactory than luminescent sheets of a solid plastic. In one test of dye stability, solutions of Coumarin 481 and 540, and Rhodamine 590 and 640 in sealed acrylic cylinders were completely bleached in five days of exposure to sunlight [17].

The luminescent concentrator is an interesting concept permitting arbitrarily high concentration ratio with flat-plate freedom from tracking. The efficiencies of known organic dyes are adequate to allow overall efficiencies (in conjunction with silicon solar cells) in the 5 to

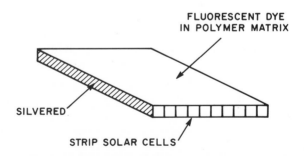

Figure 5.9   Fluorescent flat-plate concentrator (after Ref. 14, this Chapter).

10 percent range, but life of these dyes is too short to consider this a practical approach in the present form.

The thermo-photovoltaic (TPV) cell offers the possibility of using some of the infra-red energy which would otherwise not be used at all. In this design, sunlight is collected in a high ratio optical concentrator and used to heat a secondary radiator at a lower temperature than the solar surface. Less radiant energy is then present at short wavelengths above the solar cell bandgap, to be partially wasted in photovoltaic conversion. It is also possible to recycle infrared energy not absorbed in the photovoltaic cell and use it to maintain the radiator temperature (see Fig. 5.10). The attraction of this approach is that an overall efficiency of electric generation can be calculated to lie in the 30-40 percent range.

A proposed configuration is sketched in Fig. 5.11 [18] . A two-stage optical concentrator consisting of a primary paraboloidal mirror and a CPC second stage brings sunlight into the entrance throat of the TPV converter. This minimizes the deleterious effects of aberrations and surface imperfections as compared to a single high ratio paraboloid. Within the TPV converter the sunlight expands and illuminates the

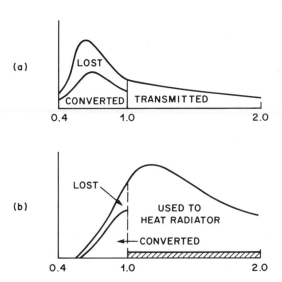

Figure 5.10   Principle of thermophotovoltaic (incandescent) converter.   (a) solar (5800K) spectrum and portions converted, wasted $(h\upsilon > E_g)$ and not used $(h\upsilon < E_g)$ for Si solar cells. (b) same for 2400K spectrum. If long-wave light is reflected back to radiator, efficiencies to 55 percent are theoretically possible.

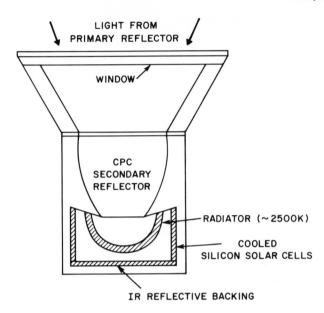

LIGHT FROM
PRIMARY REFLECTOR

WINDOW

CPC
SECONDARY
REFLECTOR

RADIATOR (~2500K)

COOLED
SILICON SOLAR CELLS

IR REFLECTIVE BACKING

Figure 5.11   Cross-sectional schematic of a TPV incandescent converter using a CPC secondary element. Cooling, electrical leads and provision for vacuum insulation are not shown (after Ref. 18, this Chapter).

black body radiator at about 300 suns intensity. Roughly 25 percent of the available energy is lost to this point because of reflector imperfections and reradiation from the black body. The secondary radiation is coupled to silicon solar cells, perhaps fabricated in two parts to give a hollow right circular cylinder and cap. (Hollow cylindrical silicon solar cells have been prepared by modified EFG growth techniques [19].) The infrared radiation not used to create election-hole pairs must be reflected back to the radiator if high overall efficiency is to be obtained, and the silicon cell must be cooled to maintain good efficiency. Absorption mechanisms not related to election-hole pair production (lattice absorption or free-carrier heating) must be small so that excessive heating of the photovoltaic cell is avoided.

Initial prototype experiments with radiators heated by electric filaments have shown efficiencies up to 10 percent. The absorption of infrared radiation has been about 15 percent and is one important limiting factor, together with lower photovoltaic efficiencies than should be obtained. The potentially serious problems posed by the need for heat-sinking and reduction of series resistance losses have been shown to be tractable, however.

5.2.3 Materials and Construction of Practical Optical Concentrators

Evidently optical concentrators for use with high intensity solar cells cannot be much more costly than flat-plate solar cell modules. Since the high concentration systems may be more efficient (perhaps 22 percent or more for systems using $Al_xGa_{1-x}As/GaAs$ cells) the DOE goals may allow for as much as $100/m^2 of aperture area for systems using concentration ratios above 500. Low ratio systems will almost certainly be similar in efficiency to flat-plate modules, and the concentrator assembly will have to cost less than the cells displaced for it to be effective.

An example of a low cost, low concentration concept is shown in Fig. 5.12(b) [22]. This type of linear, nonimaging reflector permits both sides of the solar cells to be illuminated, allowing half the number of cells to be employed. Because of the long optical absorption length in silicon, the light from front and rear surfaces overlaps in cells of 50 to 100μm thickness, giving some enhancement of V_{oc} as compared with one sun conditions. For conditions where the cost of silicon cells dominates module cost, this simple approach gives a factor of nearly two in cost reduction.

On sites where shading is not a problem, such as on the roof or sloping wall of an isolated structure, a planar reflector may be positioned as in Fig. 5.12(a) to augment the illumination of a flat-plate module. This can result in an enhancement factor in the 1.5 to 1.8 range at very low cost; but the nearly horizontal orientation of the reflector renders it vulnerable to soiling and deterioration. The real cost of this simple reflector system is likely to be about the same as that of flat-plate module encapsulants, i.e., $1.00 to 2.00/ft^2, adding $0.10 to $0.20/W_{pe}$ to the cost of the solar cells. If technology achieves the $0.50/W_{pe}$ goal for flat-plate modules, use of reflectors of this type should reduce the cost into the $0.30 to $0.40/W_{pe}$ range. These flat reflectors have essentially the same acceptance angle as flat plate arrays. Diurnal tracking is not essential, but seasonal adjustment of inclination of either the reflector or cell plane or both is beneficial.

At the next level of complexity are line-focusing imaging and nonimaging concentrators and single stage CPC concentrators. In the low concentration range (2 to 10 ×) these would be appropriately coupled with inexpensive cells similar to those contemplated for flat-plate use. The primary reflector for all concentrator systems must be made of inexpensive materials, since its large area is likely to make it dominate the overall cost. Present cost estimates [23] for parabolic trough reflector materials are listed in Table 5.2; note that *both* a reflector

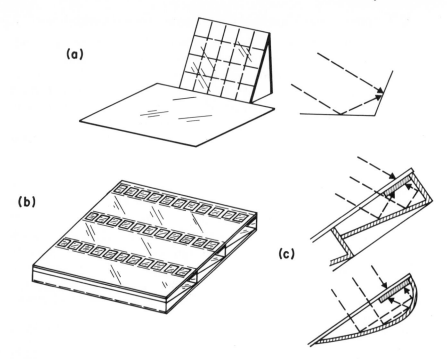

Figure 5.12   Low ratio concentrators. (a) Simple reflector augmentation, best for winter sun or high latitude use. (b) Schemes to illuminate both sides of silicon solar cells (after Ref. 22, this Chapter). (c) Cross-section of (b), with flat mirrors (M) or snail shell spiral reflector (R).

material and a structure material must be used, although the reflector price estimate includes allowance for bonding to the structure. These materials would also be suitable for linear CPC elements, but that does not seem to be a favored design for primary reflectors because the reflector area, even with truncation, is several times larger than the aperture area.

An even greater acceptance angle for the same degree of concentration is afforded by the all-dielectric CPC [24] which uses total internal reflection to concentrate the light. These could be cast of acrylic plastic, for instance, which has a refractive index $\sim$1.5 and shows good resistance to ultra-violet light. The volume of acrylic needed per m^2 of aperture ($\sim$10^{5}cm^3/m^2 for 10 cm wide $\times$ 100 cm long troughs with 4 to 5 times concentration and an 8° acceptance angle) seems excessive for economical production, however.

In the medium concentration range ($\sim$10 to 75 $\times$) more expensive cells specifically designed for concentrator application can be justified.

Table 5.2
Present cost estimates[a] for primary reflector
materials, fabricated to parabolic trough shape

Structural material	est. cost ($/ft^2)
plywood (19 mm)	1.40
fiberglass (skin + 50mm corrugated core)	2.10
Paper (skin and 50mm corrugated core)	1.90
Aluminum skin and 38mm honeycomb core.	2.50

Surface — films and foils material	est. cost ($/ft^2)
second surface glass (silvered)	2.13
metallized plastic films	0.50-1.50
aluminized mylar[b]	0.20
polished aluminum	1.50

[a] data from Ref. 22, this Chapter.
[b] requires protective bubble cover.

Linear arrays of these cells may conveniently be mounted on two adjacent sides of square or triangular cross sectioned tubing, through which coolant is circulated. Electrical connections are taken out one or both ends. Lightweight parabolic trough reflectors of several m^2 area can be made which permit tracking about an equatorial axis on a turret mount [25]. Alternatively, relatively large fields of parabolic reflectors may be rotated as a group about the vertical axis and tilt-adjusted by a coupled truss mount for elevation tracking. A proposed 10 kW$_{pe}$ array using this type of mounting is planned to operate with ¼ hp dc electric motors (1 hp = 745 watts) for azimuth and elevation [26]. At full drive less than 5 percent of the peak rated output would go to tracking. The average tracking power consumed by a properly designed system will be less than or on the order of 1 percent of the rated output.

At the high concentration level (75 to 200 suns) silicon cells are still the preferred choice but an improvement in overall efficiency into the 15 to 20 percent range can permit somewhat increased costs for more rugged and more precise reflectors and more complex cell mounts. Point-focusing optical elements requiring two-dimensional tracking of the sun are necessary. These may be completely modular, with one photovoltaic cell paired with each optical element, or they may employ a field of reflectors focused on a central array of cells (Fig. 5.13). The

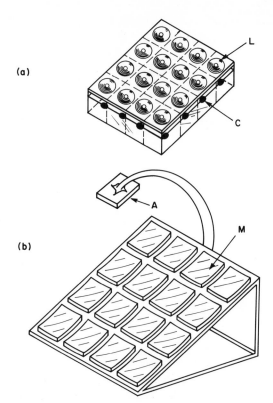

Figure 5.13   (a) Modular concentrator system. Each silicon cell (C) is illuminated by its own Fresnel lens. (b) Field of mirrors illuminating a central array of concentrator cells.

individual modules might employ, for instance, a 1.5 ft^2 acrylic Fresnel lens and a 2-inch diameter silicon concentrator cell in a liquid-cooled mount. An interesting concept making use of the fabrication technology employed in production of sealed-beam headlamps has also been described (Fig. 5.14), which offers clear advantages in terms of encapsulation integrity and soil resistance [27].

A Fresnel lens array using silicon cells of nominal 1 kW$_{pe}$ capacity has been tested for two years [28] and results indicate the drop-off in performance with time is similar to that of flat-plate modules. Sufficient information is not yet available to assure a 20-year life for acrylic Fresnel lenses, but it is possible to laminate thin plastic Fresnel sheets to the inside of glass cover sheets. Just as with flat-plate modules, an all-glass, truly hermetically sealed module is desirable.

The central receiver array offers simplified connection and cooling of the photovoltaic cells and larger optical elements (3 by 15 feet or more)

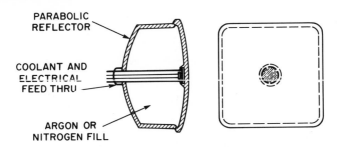

PARABOLIC
REFLECTOR

COOLANT AND
ELECTRICAL
FEED THRU

ARGON OR
NITROGEN FILL

Figure 5.14 Headlamp-type concentrator module (after Ref. 27, this Chapter).

can be used. This entails somewhat less modularity on the other hand and the advantage of plug-in repair as offered by a frame of headlight type modules is sacrificed. A higher degree of component reliability will thus be required, particularly for the photovoltaic cell assembly. Even total obscuration of one mirror panel results in only a small reduction in output, however.

At very high concentrations (500-1000 and up) GaAs based cells and spectrum-splitting high performance combination cells (Fig. 5.15) may become of practical interest. With efficiencies in the 20 to 40 percent range more rigid reflector or refractor structures can be accommodated, compatible with the need for less than $\sim 10^{-3}$ rad tracking error. The cost of dichroic beam splitters needed for separation of the solar spectrum to discrete cells is high and a tandem configuration will probably be necessary for this to become practical. In such an arrangement sunlight would be incident on a front cell of bandgap E_{G1} in which the shorter wave-length light would be absorbed with the longer wavelength photons passing into a second cell of bandgap $E_{G2} < E_{G1}$. The optimum choice of material combinations for such tandem cells has been the subject of several recent papers [29] but no experimental results either for optically stacked discrete cells or heteroepitaxially grown integrated cells have been reported. Such integrated structures would be series-connected internally and the material properties would have to be chosen so that equal photocurrents are generated if the maximum benefit were to be derived. This could only be achieved for *one* spectral distribution; other AM values would result in less than the optimum output.

Very high concentration modules could be made with similar materials as those used for 200× concentrators, but somewhat closer tolerances would be required. The high thermal and electric current loads present additional problems but these may be partially offset by the fact that waste heat may be extracted at a higher and hence more useful

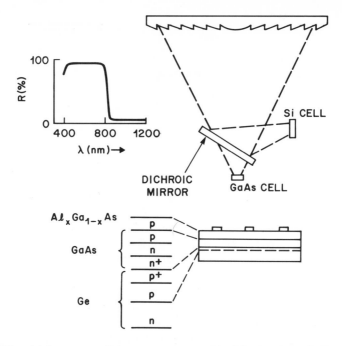

Figure 5.15   (a) Spectrum splitting concentrator with dichroic mirror (reflectivity spectrum inset).   (b) Tandem concentrator cell with wide-band-gap cell stacked above lower gap cell. The example shown could be grown epitaxially owing to the good lattice match between Ge, GaAs and AlAs; the n^+p^+ tunnel junction provides nonrectifying contact.

temperature. Using GaAs cells coolant temperatures to 200C can be tolerated, compared to about 70C which is the practical upper limit for Si based systems.

As concentration is increased, less and less diffuse sunlight is utilized. In many areas diffuse sunlight constitutes 20 to 40 percent of the total insolation and the potentially higher efficiency of concentrator systems vis-a-vis flat-plate modules is largely offset. In these locations concentrator systems must compete directly on a $/m^2 basis with the simpler flat-plate or low-concentration CPC arrays. At least one recent study suggests this is not possible [30]; that high concentration systems must offer at least 25 to 30 percent efficiency to be a viable option. If that is true, practical high concentration systems would have to await development of a tandem spectrum-splitting cell, and would even then be attractive only in desert and mountainous areas where the proportion of direct to diffuse illumination is high.

5.3 Storage and Power Conditioning

Photovoltaic electricity is available only when the sun is shining. A practical photovoltaic system must include some provision to satisfy a customer's needs at other times. Evidently 24-hour storage is a minimum, to account for nighttime needs, and at the other extreme, 180 day storage should be adequate to carry over seasonal surplus to periods of reduced supply or increased need.

In addition to the temporal mismatch between the output of photovoltaic arrays and electric load requirements, there is typically an impedance mismatch as well. The output of a photovoltaic array is direct current, proportional to the instantaneous insolation, at a nearly constant voltage. The output impedance of the array considered as a generator then depends on insolation, and it is not possible to obtain maximum power coupling to a load of any fixed impedance. Conversion of the direct current to alternating current may be required to permit compatibility with existing equipment or to permit interconnection with the conventional power grid if a supplemental rather than stand-alone system is called for. Dc to dc converters or dc to ac converters are thus required, depending on the load and storage options.

5.3.1 Storage

Lead-acid storage batteries constitute the only commercially available option for on-site storage of photovoltaic electric energy. The present cost is $\sim$ \$50/kWhr, and in deep-discharge operation a useful lifetime of about 10^3 charge-discharge cycles may be expected. Within the past year low maintenance (nonvented) batteries and no-maintenance (completely sealed) batteries have been introduced to the mass automotive battery market [31]. Clearly this is an encouraging development for the photovoltaic storage picture, but the cost is still high and one would hope for at least a ten-year life-in-service.

It appears that a minimum of 7 to 8 days storage is still a good lower bound estimate for a viable stand-alone system [32]. A 10 kW_{pe} residential array producing 40 kWhrs/day would then need about \$16,000 worth of lead acid batteries, or roughly three times the array value at the DOE 1986 price goal. Recourse to generator or utility back-up during occasional extreme conditions would still be necessary. Evidently a substantial improvement in rechargeable battery technology or the development of a cheaper alternative will have to accompany the reduction in photovoltaic array cost for stand-alone systems to become practical (see Chap. 6).

An alternative to a completely stand-alone system would be a supplemented, or fuel-saver arrangement. In the simplest case such a system would include a diesel motor-generator, standby fuel supply, reduced rechargeable battery storage, say (24 to 48 hrs) and a photovoltaic array. Under conditions where the capital cost of diesel generators is low and the cost of fuel high such a partial photovoltaic system becomes cost effective. The specific allocation to array, storage and back-up which is most advantageous depends on fuel costs, capital costs and seasonal and daily load and insolation variations.

It has been argued that the storage difficulties which stand-alone photovoltaic systems (and other solar energy systems as well) must face are economically dominant, and that solar energy will accordingly never be practical, save in those limited applications where unusual load characteristics (such as crop irrigation pumps) result in built-in storage. These arguments neglect the simplest and most direct fuel-saver mode of operation in a developed country, where the central power grid provides a convenient and readily available back-up system. In a large part of the United States electric utility loads peak during the summer and in daylight hours, so that systems with no on-site storage whatsoever may be practical. Buy-back of surplus noon-hour photovoltaic electric power by the utility would result, from the utility's point of view, in fuel saving and possibly some capital saving in peak-load generation capacity. There are a variety of complex questions to be answered before integration of privately owned photovoltaic arrays with the public (i.e., investor or publicly owned, publicly regulated) utilities can be considered acceptable, but these are not primarily technological questions. The electronic devices to interface photovoltaic arrays to the power grid exist, are efficient ($>95\%$), and are not overly expensive.

5.3.2 DC to DC and DC to AC Conversion

The variable amplitude direct current and voltage from a photovoltaic array must be converted to the form of electric power necessitated by the load requirements. For example, a constant-voltage, variable current dc supply is needed for battery charging; a constant frequency, constant amplitude ac supply is needed for operation of inductive motor loads, and phase control of the latter is also required for utility interfacing. Ac motors of modern design operate at higher efficiency than dc motors, and do not require brushes and associated maintenance. In most cases conversion to ac power is desirable, and this can be obtained with a solid-state inverter circuit using thyristor switching.

A synchronous inverter circuit resembles the familiar bridge rectifier with two important differences. The rectifier diodes are replaced by switching thyristors, and the polarity of the dc connection is reversed relative to that in an ac to dc converter (see Fig. 5.16). The switching signals applied to the thyristor control leads are derived from the utility line interface assuring synchronous phase operation. The synchronous inverter injects a current pulse back into the utility line which commences when the thyristors are turned on by the control signal and shuts off when the instantaneous ac voltage exceeds the dc supply level, reverse biasing the thyristors. This type of operation is called line-switched because the thyristor *switch-off* is obtained by this reverse bias action. Provision must be made to ensure external removal of the ac source if the ac line power fails since the thyristors would then lock on and connect dc power directly to the main.

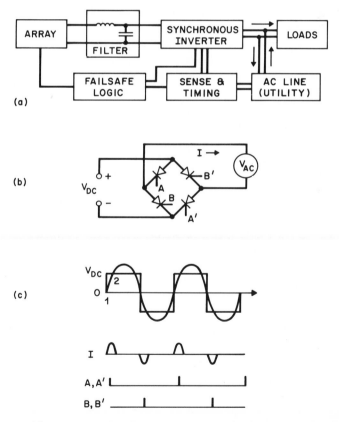

Figure 5.16 (a) Block diagram of a line-switched synchronous inverter circuit for a photovoltaic array interfaced to a utility line. (b) SCR bridge inverter circuit. Timing circuit (not shown) switches A,A' or B,B' pairs alternately. (c) Voltage and current waveforms as a function of time.

The simple line-switched thyristor inverter circuit also requires that the photovoltaic source be floated relative to ground. An isolation transformer may be provided if this condition is impractical, or if the array cannot be designed to provide an appropriate voltage match to the ac amplitude. Even harmonic distortion in the output power is low from a properly operated device, depending on the degree to which the forward and reverse conduction duty cycles in the thyristor bridge are matched. Odd harmonic distortion in the voltage waveform can be minimized by appropriate filtering of the dc supply so that it appears to have a much higher impedance than the utility line. The impedance which the inverter and line combination present to the photovoltaic array may be controlled by variation of the conduction duty cycle of the thyristors, so that the maximum power possible is converted to ac. Under these conditions the net power flow from the utility may be reduced or even reversed, depending on the local load demand.

Simple synchronous inverter equipment suitable for low to medium power applications (2 to 100 kW) which operates essentially as described is commercially available [33]. More complex, sophisticated circuitry may be used to provide reduced harmonic distortion or load management functions; these are accomplished by using power transistor bridges (rather than thyristors) and microprocessor controlled switching sequences. Appropriate sensors must be included to insure disconnect from the line and/or load if the array or line voltages deviate outside design limits, or if other operation error is present. Transistorized bridge inverters need not be line switched and accordingly an output very low in harmonic content can be obtained [34]. Operation in a stand-alone mode apart from a utility line is also possible, but an initial dc to dc converter stage must be added to stabilize the variable array voltage to a fixed amplitude prior to inversion.

Dc to dc converters may be designed for either step-up or step-down operation or for up and down conversion near a unity ratio. The conversion ratios may be continuously varied by control of the duty cycle of switching elements (power transistors, for example). These converters contain inductive and capacitive elements, as well (see Fig. 5.17). In battery charging applications a dc to dc converter may also serve as a maximum power tracker, effectively presenting an optimum load impedance to the photovoltaic array to insure maximum utilization of the available power regardless of variations in temperature or insolation. Present efficiencies for dc to dc conversion are in the 95 to 98 percent range. The challenge is one of optimizing circuit layout to minimize bulk and cost of the reactive circuit elements required.

(a)

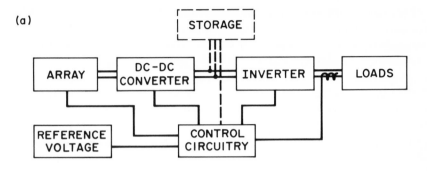

(b)

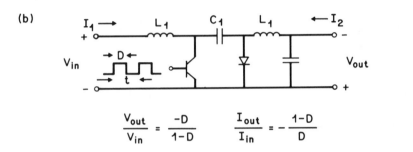

$$\frac{V_{out}}{V_{in}} = \frac{-D}{1-D} \qquad \frac{I_{out}}{I_{in}} = -\frac{1-D}{D}$$

Figure 5.17 (a) Block diagram of a self-switched synchronous inverter in a stand-alone ac/photovoltaic system. (b) Dc-to-dc converter circuit. The output voltage may be raised or lowered by varying the duty cycle of the transistor switch.

5.4 Endurance Testing, Standards and Certification

Customers for photovoltaic electric systems must be persuaded that they are reliable and cost-effective, both in the direct sense that maintenance or repair requirements can be predicted, and in the indirect sense that electric power will be available as needed so that the expenses of production shutdown or cost of backup power purchase will not be prohibitive. Credibility and confidence cannot be established quickly or by breakthroughs in technology. At the present time there is only limited information available of the actual in-field performance of photovoltaic systems, in part because of the limited number and recent establishment of test sites and field trial installations, and in part because of incomplete instrumentation, data gathering, and data correlation. To date, field studies have for the most part been directed toward the establishment of initial performance description rather than life projection.

Life-testing in the midst of technological evolution is difficult and poses the danger of discouraging the very changes that might have greatest impact on lifetime. This comes about because life-testing is essentially a statistical process of failure analysis. One has no information without the actual occurrence of failures, and the significance of that information increases in proportion to the number of identical elements under test. Further, establishment of a minimum probable lifetime in real-time testing takes a period at least equal to that lifetime, and hence as a practical matter accelerated and/or curtailed test methods must be employed. Unfortunately accelerated testing methods require rather detailed knowledge of the types and mechanisms of failure, a knowledge which only real-time experience can provide. Absent those results, assumptions must be made, or a model assumed; for example, that modules will fail at twice the rate or in half the time from humidity related effects if the relative humidity is doubled, or that failure will occur at an exponentially faster rate as temperature, say, is increased. An actual solar cell module can fail for a variety of reasons, since failure only means that output has dropped below some minimum acceptable level. For modules containing single crystal silicon cells, the important failure modes have been shown to be related to contact and interconnection corrosion, loss of cover transparency, or physical fracture of the cell material. These modes appear to reflect faulty workmanship or improper choice of encapsulant rather than an intrinsic phenomenon, such as aging of junction properties. Vandalism or physical injury from natural events may initiate a new gradual degradation mode in addition to the catastrophic failures one would expect from such external causes. Lifetime projection thus should include some idea of site security as well as the effectiveness of module encapsulants in resisting projectile penetration as well as humidity and air pollutants.

Concentrator arrays are exposed to all the lifetime limiting factors which affect flat-plate modules and suffer from additional complications of reflector and tracking performance. Unless the cells are operated at deliberately elevated temperatures, cell reliability problems should be no worse and probably somewhat better than in flat-plate arrays, because there are many fewer cells and overall reliability increases rapidly with a decrease in the number of components subject to failure. Concentrator cells can be more expensive and hence additional quality control can be employed to eliminate workmanship errors, as well. Concentrator arrays are much more sensitive to optical deterioration than flat-plate modules, however. The aluminized plastic films which permit inexpensive concentrator construction are sensitive to chemical air pollutants as well as particulate abrasion. Some of the lightweight structural materials, such as aluminum sheet, for example, are readily deformed by hail.

Standards for manufactured products consist of certain test specifications, and certification means that either a particular product or fully representative samples of a quantity of that product have performed to pass the test specifications. Obviously standards must be established with some insight into failure mechanisms if certification is to imply anything about projected lifetime, but they can serve to assure freedom from initial defects which may cause subsequent failure. At the same time unnecessarily stringent adherence to standards may increase the cost of systems and in worst cases may actually contribute to locking in long term failure modes, since compliance with standards reduces innovative trial.

Considerable experience has been developed for product certification, standards definition, and reliability testing in the areas of military systems, space systems, and telephone communication equipment. Each area tends to emphasize different criteria for acceptability. The military standards tend to emphasize accelerated torture testing, the space program emphasizes nondestructive tests and quality control inspections, and the telephone equipment industry is oriented toward moderately accelerated life studies under specific environmental conditions. Reliability tests for photovoltaic components and systems would have to be somewhat different in approach, to account for the facts that the use environment is neither so well controlled as in a modern telephone exchange, nor so unpredictable as in a theater of military operations, for example. Perfect no-maintenance performance is not an essential feature of photovoltaic systems; while it is the only acceptable level for space probes or manned space vehicles.

Such photovoltaic installations as are now in place are better suited for initial evaluation tests than life or aging tests. Some interesting observations can be made, however, in spite of the brief experience available flat-plate panels appeared in practice to be as robust in performance and as resistant to weather and soiling problems as one might have optimistically expected. During January of 1978 a flat-plate array at the Lewis Research Center (NASA) in Cleveland was exposed to 82 mph winds and driving snow during one of the worst blizzards ever recorded in that area. No damage was incurred, and the array output remained at about 20 percent of normal peak power during the storm [35]. An array installed at the summit of Mt. Washington, NH, survived encasement in rime ice for over two months as well as winds to 133 mph with only minor surface damage from windborne ice projectiles.

The photovoltaic modules at the Mead Field Station of the University of Nebraska (irrigation and crop drying system) have fared less well. There are two types of silicone encapsulated modules there; over half of the modules of one type have experienced cracked cells over the July, 1977 - September, 1979 period. The number of cracked cells is only ~2.2 percent of the total, however — most resulting from hail impact. The other cell failure mode, also evidenced by cracking, resulted from thermal mismatch and stress ocurring in one of the module designs. A cell cracked completely through will cause electrical failure of its series string (or module, for series wired modules). Many of the hail-impact cracks did not cause module failure. Indeed only 48 of the total 2250 modules failed although 1044 of the 2080 modules inspected contained one or more cracked cells.

Another large installation is the photovoltaic residential power experiment at the University of Texas at Arlington, initiated in November of 1978. After 7 months of highly successful operation the system was sut down for module cleaning and inspection and the modules were placed in a short-circuit mode. This appears to have caused overheating, delamination and cracking due to thermal stresses. During the next several months over one-quarter of the modules were found to have failed, and it was concluded that these failures were due to reverse bias voltages appearing across some cells under the short circuit, illuminated conditions associated with the washing and inspection procedure. The modules at Mead have been subjected to similar treatment many times without incident, however, so this appears to be a characteristic of a particular module design. The modules under evaluation at Mead and Arlington were from the DOE Block II purchase (1977) and do not incorporate improvements in module design arising in the last 3 years [36].

Concentrator arrays have not been in place long enough for even such anecdotal information to be collected. Several low power arrays were installed at Sandia Laboratories in 1977. Even in the relatively moderate New Mexico environment encapsulation failure and other initial design defects have dominated the early experience. These have been corrected and the test facility there is being expanded to include testing of power conditioning equipment as well as concentrator modules of various types [37].

Lightning protection is an additional area of concern. The technology for almost any desired degree of protection exists, but a sensible trade-off between estimated risk and cost of protective equipment must be made. An initial study bears out the expectation that voltage transients associated with nearby vertical strikes pose the major danger [38]. No incidents of module failure due to lightning strikes have been reported to date, however.

5.5 Cost Impact of BOS Elements

The 1986 DOE goal of $0.50/W_{pe}$ for photovoltaic modules represents only half the cost picture. The cost of other system elements, or the balance-of-system (BOS) costs, are supposed to be comparable. Included in BOS costs are power conditioning, storage (if any), site preparation, structural supports, electric wiring, installation labor, and test and acceptance check-out expenses. At the present time, BOS costs actually *exceed* the cost of flat-plate modules on a $/W_{pe}$ system basis. This is particularly disturbing as most BOS items do not appear to be amenable to cost reduction through economies of scale. The cost of wiring or lead-acid batteries is materials-dominated and the battery and wire manufacturers are already in large-scale production. Land costs will certainly not decrease. Perhaps the cost of structural supports could be reduced by standardized mass production. The use of photovoltaic roofing materials in new construction would appear to allow a substantial effective saving from the displacement of conventional roofing materials and installation labor.

Estimated present and projected prices for various BOS elements are summarized in Table 5.3.

Table 5.3
Present and Projected BOS costs (per W_{pe}) for
residential-scale flat-plate systems

Size	2kW	$5-10kW_{pe}$	$5-10kW_{pe}$
System	free-standing	on roof	as roof
Time	1978[a]	projected[b]	projected[c]
Site & Structure	2.40	0	−.05
Wiring	3.0.0	0.08	0.04
Power Conditioning	0.50	0.20	0.20
Installation	2.00	0.40	0.07
Storage (optional)	(2.0)	(0.50)	(0.50)
Total	$8. + (2.)[d]	0.68(+0.50)	0.25(+0.50)

[a] from Ref. 37.
[b,c] from Ref. 35
[d] actual range (from Ref. 37) of prices quoted by manufacturers in 1978 — $8.63 to 13.88 (1975 dollars).

Optimistically, it appears that a goal range of $0.30 to $0.70/$W_{pe}$ may be appropriate either for residential scale systems *without* storage and with full roofing credit *or* for large scale power plant use, again without storage [39]. Lead-acid battery storage for a nominal 3 kWhr/kW_{pe} rating would double or triple the BOS costs, assuming amortization over a 20-year array lifetime, a 10-year battery life and a $50/kWhr price. Storage at this price could only be justified in areas of high average insolation (e.g., the American southwest) and only if utility buy-back was not allowed. Even at one-third the retail price for consumer electric power, sell-back to a cooperative utility is less expensive for the consumer than battery storage at this price level.

If one considers the *actual* present BOS costs for delivered, installed arrays in DOE funded programs, the picture is even more gloomy. Cost analyses for 1978 indicate a BOS cost range of $8.60 to $13.90/$W_{pe}$ [40], as compared to a range of $6 to $8 for concentrator modules and $9 to $13 for flat plate purchases. Even if BOS costs can be cut in half by the year 1986, 90 percent of the cost of a photovoltaic system would arise from elements other than the solar cell modules at $0.50/$W_{pe}$. BOS costs other than storage are largely independent of the actual power produced by flat-plat arrays, tending to vary instead with system area [41]. Thus advances in *efficiency* are likely to be more important than advances in cost/W_{pe} of the photovoltaic modules. This accounts for much of the skepticism in the engineering community regarding the potential impact of promised breakthroughs in thin-film solar cell materials technology. A reduction in storage costs by an order of magnitude, however, *would* be significant. In the absence of such a reduction any significant utilization of photovoltaic systems in developed countries will require integration with the established utility power grid or use with intrinsically storing loads such as water pumping or warehouse refrigeration which can accept intermittent energy input.

Overall one must conclude that just as solar module costs must be reduced by a factor of 10 to meet the DOE goals, the cost of the other major components of a practical photovoltaic system must *also* be reduced by a similar factor. Electrical wiring, site preparation, and installation labor will be particularly difficult areas for cost reduction and do not seem likely to benefit from new technology. The cost of lead-acid batteries is largely cost-of-materials (lead) dominated, and no practical alternative is visible on the technological horizon. Indeed, cost projections for other advanced energy storage systems such as high performance flywheels [42] suggest a possible future price range essentially the same as that which today's battery technology already affords. Power conversion equipment, already one of the cheapest of the essential BOS elements, seems the only area where cost reduction with somewhat improved technology and increased volume is predictable. Fortunately this should provide engineering leeway to assure a credibly safe, acceptable utility interface which seems essential for near-term, cost-effective photovoltaic implementation in the United States.

REFERENCES

1. D. M. Chapin, *Proc. 1955 Conf. on Use of Solar Energy*, Vol. 5, 117 (U. Ariz. Press, Tucson, AZ, 1958).

2. N. F. Shepard, Jr. and L. E. Sanchez, *Proc. 13 PVSC*, 160 (IEEE New York, NY, 1978).

3. G. Wallis, *J. Amer. Ceram. Soc.* **53**, 563 (1970).

4. P. R. Younger, W. S. Kreisman, G. A. Landis, A. R. Kirkpatrick and R. F. Holtze, *Proc. 13 PVSC* , 729 (IEEE, New York, NY, 1978).

5. J. A. Minucci, A. R. Kirkpatrick and W. S. Kreisman, *Proc. 12 PVSC*, 309 (IEEE New York, NY, 1977).

6. SES, Inc. Newark, DL.

7. E. Anagnostou and A. F. Forestieri, *Proc. 13 PVSC*, 843 (IEEE New York, NY 1978).

8. G. B. Gaines, D. C. Carmichael, F. A. Sliemers, M. C. Brockway, A. R. Bunk, and G. P. Nance. *Proc. 13 PVSC*, 615 (IEEE New York, NY, 1978).

9. N. F. Shepard, Jr. and L. E. Sanchez, *Proc. 13 PVSC*, 160 (IEEE New York, NY, 1978).

10. L. W. James and J. K. Williams, *Proc. 13 PVSC* , 673 (IEEE New York, NY, 1978).

11. A general treatment of low-ratio concentrators appears in the recently published *Optics of Nonimaging Concentrators: Light and Solar Energy* by W. T. Welford and R. Winston (Academic Press, New York, NY, 1978).

12. R. Winston, *Solar Energy* **16**, 89 (1974).

13. A. Rabl, *Solar Energy* **18**, 93 (1976).

14. W. H. Weber and J. Lambe, *Appl. Optics* **15**, 2299 (1976).

15. B. D. Wedlock, *Proc. IEEE* **51**, 694 (1963).

16. C. F. Rapp and N. L. Boling, *Proc. 13 PVSC*, 690 (IEEE New York, NY,, 1978).

17. J. R. Wood and J. F. Long, *Proc. 13 PVSC*, 1158 (IEEE New York, NY,, 1978).

18. R. N. Bracewell and R. M. Swanson, *Report ER-633*, Electric Power Research Institute (EPRI), Palo Alto, California (1978).

19. A. I. Mlavsky, H. B. Serreze, R. W. Stormont and A. S. Taylor, *Proc. 12 PVSC*, 160 (IEEE, New York, NY, 1977).

20. E. D. Jackson, *Proc. 1955 Conf. on Use of Solar Energy*, Vol. 5, 122 (U. Ariz. Press, Tucson, AZ, 1958), also E. D. Jackson, *U. S. Pat #2,949,498* (1960).

21. R. L. Moon, L. W. James, H. A. Vander Plas, T. O. Yep, G. A. Antypas and Y. Chai, *Proc. 13 PVSC,* 859 (IEEE New York, NY, 1978).

22. Y. Chevalier, F. Duenas and I. Chambouleyron, *Proc. 13 PVSC,* 738 (IEEE New York, NY, 1978).

23. M. W. Edenburn, D. G. Schueler and E. C. Boes, *Proc. 13 PVSC,* 1029 (IEEE New York, NY, 1978).

24. N. B. Goodman, R. Ignatius, L. Wharton and R. Winston, *Appl. Optics* **15,** 2434 (1976).

25. Solarex Corporation, Rockville, Maryland, *Application Note 6906* (1978).

26. J. A. Castle and K. Romney, *Proc. 13 PVSC,* 1131 (IEEE, New York, NY,, 1978).

27. W. Masters, R. Maraschin, and W. Kennedy, *Proc. 13 PVSC,* 686 (IEEE, New York, NY, 1978).

28. E. L. Burgess and D. A. Pritchard, *Proc. 13 PVSC* 1121, (IEEE, New York, NY, 1978).

29. In addition to Moon, et al. (Ref. 21) cf. A. Bennett and L. C. Olsen, *Proc. 13 PVSC,* 868 (IEEE, New York, NY, 1978); J. A. Cape, J. S. Harris, Jr. and R. Sahai, *ibid,* 881; M. F. Lamonte and D. Abbott, *ibid,* 874.

30. E. A. DeMeo and P. B. Bos, *Report ER-589-SR,* Electric Power Research Institute (EPRI), Palo Alto, California (1978, unpubl.).

31. For a recent review of secondary battery technology, see A. J. Salkind, D. T. Ferrell, Jr. and A. J. Hedges, *J. Electrochem. Soc.* **125,** 311C (1978).

32. A. S. Barker, Jr. Priv. Comm. (1978).

33. For example, Windworks, Inc., Mukwonago, WI, USA manufactures synchronous inverters in the 1 kW to 1 MW range.

34. S. Cuk and R. D. Middlebrook, *Proc. 1977 Power Electronics Specialist Conference,* 160 (IEEE, New York, NY, 1977).

35. R. C. Cull and A. F. Forestieri, *Proc. 13 PVSC,* 22 (IEEE, New York, NY, (1978).

36. S. E. Forman and M. P. Themelis, *Proc. 14 PVSC* (to be publ., IEEE, New York, NY, 1980).

37. J. L. Watkins and D. A. Pritchard, *Proc 13 PVSC,* 53 (IEEE, New York, NY, 1978).

38. C. B. Rodgers, *Proc. 14 PVSC* (to be publ., IEEE New York, NY, 1980).

39. G. J. Jones and D. G. Schueler, *Proc. 13 PVSC,* 1160 (IEEE, New York, NY, 1978).

40. G. F. Hein, J. P. Cusick and W. A. Poley, *Proc. 13 PVSC,* 930 (IEEE New York, NY, 1978).

41. E. L. Burgess, *Proc. 14 PVSC* (to be publ., IEEE New York, NY, 1980).

42. A. R. Milner, *Technology Review* (November, 1979) pp 32-49.

CHAPTER 6

Economic and Social Considerations

In the preceding chapters we have discussed the science and technology of photovoltaic devices and systems, emphasizing the areas where progress toward the DOE cost and efficiency goals seems possible. The implicit assumption has been, that if these goals were met, a substantial solar photovoltaic contribution to our energy needs would follow. That does not necessarily have to be the case as there are a number of other considerations which go beyond purely scientific or even engineering discussion. These include the effects of tax credits or penalties, fluctuating interest and inflation rates, cost escalation of competitive energy supplies, and impact upon the environment.

At best these effects can be studied quantitatively in terms of postulated models, which allow the influence of changes in interest and inflation rates, for example, to be computed and an optimum photovoltaic or hybrid photovoltaic/solar heating configuration to be established *under certain assumptions* of insolation, capital cost of array and storage, maintenance cost and interval, competitive energy costs, tax, interest, and inflation rates, and anticipated load characteristics. Only some of these factors are known ahead of time — the efficiency and areal cost of the array, the installation cost, and the expected average insolation are numbers which will not change after a decision to go solar is made. The interest rate may or may not change thereafter, depending on the financing terms, while taxes, inflation, and competitive fuel costs will almost certainly vary, perhaps substantially, over the 20 to 30 year life-time expected for a solar photovoltaic installation. If storage is made up in part or in whole by sell-back to the utility grid, then the terms on which that is contracted become important as well.

During the past decade there has been an increasing popular aware-ness that the habitable environment of the earth has been substantially abused during the evolution of today's industrial and technological society. This awareness has been reflected in regulatory actions limiting the development of energy sources, particularly refusals to issue per-mits for new nuclear, hydroelectric and fossil fuel powered electric generating plants, and restrictions on the strip-mining of coal. The wide-scale implementation of solar photovoltaic generation will certainly have an environmental impact of its own, and that must be assessed and compared to the expected or known impact of other options.

Economic analyses and environmental impact assessments do not, unfortunately, predict the future in an absolute way. At best they can serve to reduce the number of options or clarify the effect of social and economic policies on a given option. In this chapter we wish to discuss, in general terms, the nontechnical factors which will determine when (or whether), how fast, and in what guise photovoltaic electric genera-tion may become a commercial reality.

6.1 Standard Economic Analyses

To illustrate the factors which affect a decision to install a photovol-taic electric system, consider first the simple example of a home builder who wishes to know whether he will save money by adding such a sys-tem to a new house. He must first determine how much the array and other components will cost, installed; then compute how much *useful* electricity will be produced (i.e., at a time when it can be used, or stored, or sold back to the utility grid), then compute the value of that electricity and hence the savings realized. Evidently if the savings over the life of the system exceed the cost it is a viable investment.

The cost of a residential photovoltaic system is almost entirely first cost. This number and the engineering data for electric production (based on array efficiency and average insolation) may be taken as fixed and known at the outset. To simplify matters, let us assume there are *no* further costs; for instance, the homeowner may wash the array him-self, periodically, and operate without on-site storage, so that no other maintenance costs are incurred. Let us assume an average load of 1200 kWhrs/month and an array sized to produce 400 kWhrs/month of power, on average, chosen to match the hypothetical requirements shown in Fig. 6.1. We assume the electricity produced can all be con-

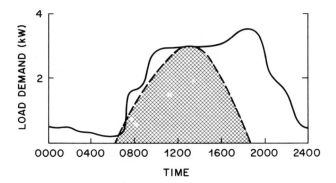

Figure 6.1 Hypothetical electric demand as a function of time of day for residential loads. The average availability of photovoltaic power from a $3\,kW_{pe}$ array is shown hatched.

sumed and hence has the full value ($.05) of the retail utility rate. For a 20 year mortgage at an interest rate of 10 percent per annum, how much of an increase in the price of a house due to the photovoltaic system can be accepted?

First, the effective interest cost is less than 10 percent because of inflation. At the end of the first year the amount (see Table 6.1 for list of symbols) of indebtedness is $S(1+r)$ but the first payment has become $P(1+i)$ so that the remaining balance is

$$S(1+r) - P(1+i) \qquad (6.1)$$

At the end of the second year, the account stands at

$$[S(1+r) - P(1+i)](1+r) - P(1+i)^2 \qquad (6.2)$$

and after N years,

$$S\,(1+r)^N - P(1+i)^N \left[1 + \frac{1+r}{1+i} + \cdots \left(\frac{1+r}{1+i} \right)^{N-1} \right] \qquad (6.3)$$

which is to equal zero when N equals the total number of payments. The bracketed term may be rewritten using the identity

$$X^N - 1 = (X-1)\,(X^{N-1} + X^{N-2} + \cdots + X + 1) \qquad (6.4)$$

Using $1+r' \equiv (1+r)/(1+i)$, (6.3) becomes

$$P = \frac{S\,r'\,(1+r')^N}{(1+r')^N - 1} \qquad (6.5)$$

Table 6.1
Symbols used in Section 6.1

S	-	net capital cost of system components and installation including land and structure costs and tax credits (if any).
i	-	annual inflation rate in general economy.
r	-	annual interest rate.
r'	-	true interest rate to lender, corrected for inflation rate.
r''	-	true interest cost to borrower, corrected for effects of inflation and tax structure.
e	-	annual rate increase in cost of electricity.
e'	-	annual rate of increase in cost of electricity corrected for inflation (escalation rate).
T	-	present value of constant annual payment for tax liability over life of system.
t_p'	-	tax rate on property value, after taking Federal tax structure into account.
t_i	-	marginal tax rate on income.
N	-	lifetime; number of years in financing agreement.
P	-	present value of constant annual payment (to repay loan S in N years at interest rate r').
E	-	present value of annual savings from net reduced purchase of utility electricity.
A	-	amortization rate, P/S.

(For monthly payments, (6.5) may be rewritten with $(1+m') = (1+r')^{1/12}$ and $'M = 12N$ in an obvious way.) P gives the constant current dollar annual payment to repay the loan S, i.e., after discounting the future effect of inflation, and is accordingly called the discounted present value of the average payment.

We also need a similar quantity describing the cost of electricity during the payback period. It is generally believed that rising fuel prices will drive the cost of electricity up at least as fast, and probably faster, than inflation in general. If the present cost of the electricity produced by the system is $\$$, at the end of the first year it is worth

$$\$ (1+e)$$

This is paid for with inflated money, so the present dollar cost is

$$\$ \left[\frac{1+e}{1+i} \right]$$

The second year the same amount of electricity is worth

$$\$ \left[\frac{1+e}{1+i}\right]^2$$

in present dollars, and over the N years term the total value is

$$\$ \sum_{j=1}^{N} \left[\frac{1+e}{1+i}\right]^{j}$$

Thus the discounted present value of the average annual savings is (again using (6.4) and introducing $1+e' = (1+e)/(1+i)$):

$$E = \frac{\$}{N} \frac{(1+e')^{N+1} - (1+e')}{e'} \qquad (6.6)$$

The condition for the investment to be profitable is $P < \overset{\text{\tiny 1}}{E}$.

For our example, $\$ = \$240.$, $N = 20$ yrs, $r = 10$ percent and we will further assume $i = 8$ percent and $e = 12$ percent. Then $r' = 1.85$ percent and $e' = 3.7$ percent, $P = 0.06S$, $E = \$342.13$ and the maximum capital investment that could be justified is $S = \$5702.00$.

To produce the assumed 400 kWhrs/mo, an array operating at 10 percent efficiency would be 30 m^2 in area and have a 3 kW$_{pe}$ rating. The *total* installed system cost would then have to be less than \$1.90 per peak watt for this system to be cost-effective. If the escalation of electric costs only equals the interest rate paid ($e=r$) then the maximum allowable capital investment becomes \$4650 and the price per peak watt \$1.55 for the installed system.

As a practical matter, an individual homeowner must consider the tax structure that applies to his investment as well as the economic factors. In the United States a private individual may deduct the cost of interest paid to a lender from his income subject to Federal income tax, but may *not* deduct the cost of electricity purchased. This provides a benefit in the form of an effectively subsidized interest rate; i.e., the taxpayer would have to pay a portion of the cost of the photovoltaic system to the government rather than the bank whether he bought the system or not. In terms of our example, if we assume the homeowner is in a marginal tax bracket of 38 percent (joint return taxable income of \$35,000.), r'' becomes -1.4 percent, $P = 0.043S$ and one could invest \$7961 (vs. \$5702) or \$6488. (vs. \$4650) for the two cases considered above. These correspond to \$2.65 and \$2.16 per W$_{pe}$ system pricing.

The other effect of taxation is to make a purchase *less* attractive. In the United States a homeowner typically pays local municipal property tax based on the market value of the real property; i.e., land and permanent improvements. A photovoltaic electric system would generally be considered a permanent improvement, like a furnace or bathroom, rather than a portable appliance, like a dishwasher; and hence would be subject to local property tax in most parts of the United States under current law. Local tax rates vary substantially, but 2 percent of market value may be taken as typical of suburban areas.

Predicting the effect of local taxes in a general way is difficult, but several points are worth making. First, they are deductible from Federally taxed income. Second, we may assume the rate is constant and that some allowance for depreciation of the value of the array will be given; for instance, we may assume a linear depreciation over the life of the array. Then if we also assume that the array is appraised at the start of each year, that taxes are paid at the end of each year, and that inflation tends to increase the market value of non-depreciated arrays at the same rate as other items in the general economy, the tax cost in the first year is

$$T_1 = t_p'S \left(\frac{1}{1+i} \right) \tag{6.7}$$

The second year tax cost is

$$T_2 = t_p'S(1+i)\left(1-\frac{1}{N}\right)\cdot\left(\frac{1}{1+i}\right)^2 \tag{6.8}$$

and the Nth year the cost is

$$T_N = t_p'S\left[(1+i)\left(1-\frac{1}{N}\right)\right]^{N-1}\cdot\left(\frac{1}{1+i}\right)^N \tag{6.9}$$

The total property tax cost including linear depreciation and inflation is then

$$\sum_{i=1}^{N} T_i = \frac{t_p'S}{1+i}\sum_{j=0}^{N-1}\left(1-\frac{1}{N}\right)^j \tag{6.10}$$

and using (6.4) the present value property tax cost *per year* is

$$T = \frac{t_p'S}{1+i}\left[1-\left(\frac{N-1}{N}\right)^N\right] \tag{6.11}$$

For the earlier example, with $i=8$ percent and $N=20$ years, a 2 percent property tax rate becomes an effective rate of 1.3 percent after allowing for Federal tax deduction, and $T=0.0043\ S$. This reduces the allowable installed system costs from \$2.65 to \$2.41 per W_{pe} for the case with $r=10$ percent and $e=12$ percent considered above. Thus the effect of a 2 percent local property tax on a homeowner is to reduce the allowable system cost by about 10 percent.

The numerical results obtained in this section serve to show that residential photovoltaic systems will become cost effective at installed prices in the $1.50 to 2.50 per peak watt range, at least for the particular assumptions made here. These assumptions were intended to be reasonable, but some comment is in order. Perhaps the weakest assumption˙is the value of electricity produced, which was based on a current rate of $0.05/kWhr. This rate is reasonably representative for 1978 retail pricing to residential customers in the northeastern United States, while rates in the $.07-.08/kWhr applied in 1979 in the NY-NJ metropolitan areas. while rates in the $.07-.08/kWhr applied in 1979 and early 1980 in NY-NJ metropolitan areas. It may be argued that electric rates in other areas are lower — 30 percent lower in the southwestern United States, for instance. In that area the lower utility rate is approximately offset by higher insolation, so the system described will produce more electricity corresponding to about the same dollar value. Thus the overall effect on breakeven pricing will not change greatly. Another serious objection is the assumption that all the electricity produced is used on-site or sold back to the utility at the full retail rate. The latter assumption is plainly unrealistic and the former limits the system size to 40 percent or less of the total load, and would still requ˛re some local load management (baking, cooking and washing during daylight hours, for instance) at the upper end of that range.

The precise value of inflation and interest rates assumed are largely unimportant since the banking system will endeavor to track them to retain an effective interest rate in the 1.5 to 2.5 percent range, which should provide an adequate real-dollar return on long-term minimum risk loans. This assumes, of course, that photovoltaic systems are sufficiently reliable and long-lived that they may be treated in the same category as dwellings. We have ignored the additional insurance costs for underwriting a photovoltaic system. This should be negligible, perhaps one-tenth of property tax cost, provided the kind of reliability that would be required for home-mortgage financing is demonstrated.

The assumed rate for escalation of conventional electricity is more critical. As this exceeds the interest rate, the attractiveness of a photovoltaic system increases rapidly. *The most substantial impediment to early utilization of photovoltaic systems would be continued government-mandated pricing of conventional electricity below its true energy-replacement cost.* By the same token, taxes or other factors increasing the cost of fossil or nuclear fuels and electric generating plants using them make early acceptance of photovoltaic systems more probable. Since the cost of fuels and generating plants is largely determined by political factors at the regional, national and international levels, rather than by technical factors amenable to rational analysis, it is difficult to say whether a fuel escalation rate in the 1.5 to 4 percent range is probable during the next few decades or not.

We have ignored direct government subsidy in the form of possible tax credits or otherwise. The state of California presently allows a credit of up to 55 percent of the capital cost of solar energy equipment to be applied by individuals against state income tax over a three-year period, while the United States government allows a one-time credit of 30 percent (up to $2200) for installation of active solar energy equipment including photovoltaic arrays. It is certainly possible that local tax credits might be granted as well. These subsidies obviously enhance the attractiveness of photovoltaic equipment to the purchaser, but they represent a cost burden to other taxpayers.

We have ignored on-site storage. Eight-day storage would add over $5000 to the system price in the above examples, roughly doubling the cost. This would require an electric cost escalation rate of more than 12 percent to justify inclusion, much higher than the 2 to 4 percent range which appears likely. A reduction in the cost of storage by roughly a factor of 10 from the present price of $50/kWhr would be necessary, together with a 20-year lifetime. There appears to be no prospect for lead acid battery technology to reach this goal.

While battery storage seems impractical for the residential system on its own, there is a potential exception provided by the development of electric automobiles. Provided the automobile is at home during daylight hours, its batteries can be charged with excess electricity. The deep-discharge batteries used in traction applications have a short storage life and need to be kept recharged.

In the absence of storage, local load management and favorable utility buy-back rates become essential. The advent of inexpensive home minicomputers at prices comparable to color TV receivers suggests that the load management aspect could be automated readily, but one must wonder whether it is realistic to expect the average American homeowner (with a fifth-grade level of mathematical literacy) to cope with programming a computer to run his or her household. Hardwired, plug-in micro-processor control is also a possibility, of course. The possibilities for favorable buy-back rates will depend on the degree of local penetration of photovoltaic systems, the timing of load-peaking relative to daylight hours and summer/winter variations, and the cost of peaking generators and displaced fuel. The availability of cooperative loads which can accommodate excess electric power during sunlight hours and can go without on cloudy days is an important factor. In general this will vary from one part of the country to another and from one utility to another, but in many instances excess generator capacity is presently available during the night and load peaks occur during daylight (and summer) hours.

It is interesting to note that without any direct subsidy, a photovoltaic system without storage becomes attractive to a private homeowner at a total installed price of about $2.00 to 2.50 per peak watt (in 1978 dollars) given the present income and property tax structure and 1978 interest, inflation and electric cost escalation rates. This system cost *should be met* when the module price goal of $0.50 per peak watt is reached. The allowance of over three times the module price for power conditioning equipment, wiring and installation, architectural and engineering fees and profit appears adequate. The initial installation of residential photovoltaic systems in significant numbers may thus be much closer to realization than is generally believed, and should occur well before the beginning of the 21st century.

6.2 Hybrid and Nonresidential Systems

A simple system consisting of a photovoltaic array and power conditioning equipment alone may not be the most cost effective if a total solar energy system is considered. Such a system would include solar thermal as well as photovoltaic collectors, or possibly hybrid collectors having both useful electric and thermal outputs. A heat pump of the conventional electric powered vapor-compression type, or an absorption air conditioner and solar thermal hot water heater, or a high performance Rankine turbine-driven vapor compression heat pump (with electric drive back-up) would be employed to satisfy heating, air-conditioning and hot water loads (see Fig. 6.2).

The analysis of such a system consists in determining the optimum allocation of capital to photovoltaic, hybrid, or thermal solar collectors, to electric and thermal storage capacity, and to heat pump type and size. This is possible only for cases where the detailed characteristics of insolation, temperature, humidity, and electric loads as a function of the time of day and day of the year are known. The cost, maintenance requirements, and lifetime of the various components must also be given, as well as thermal or electrical efficiency at various temperatures. For example, lead-acid batteries have greater storage capacity at moderately elevated temperatures than below room temperature, while photovoltaic array efficiency is enhanced at reduced temperatures. Calculation of the system cost-effectiveness requires an assumption of discount and energy escalation rates. Evidently a detailed analysis is a

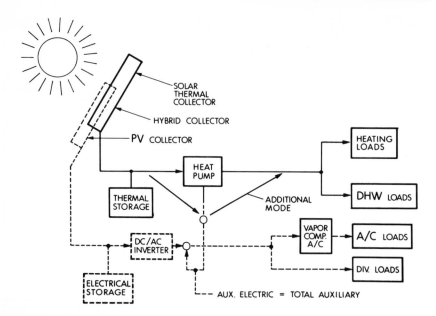

Figure 6.2 Hybrid solar total energy system, providing domestic hot water (DHW), air conditioning (AC) and space heating, and electric power to diversified (DIV) lighting and appliance loads. The compressor in the advanced heat pump may be driven from electric or thermal (Rankine turbine) power. (Reprinted from Ref. 1, this Chapter, with permission of the authors.)

highly complex task, even though the same discounted present value approach used in the preceding section can be applied.

There is an important difference between business and residential applications which arises from different accounting practices. The purchase of fuel or conventional energy is considered a normal cost of doing business, and is paid from current cash flow. The purchase of photovoltaic or solar thermal equipment, however, would be capitalized; i.e., financed by borrowing, or issuing stock, or both. Approximately half the cost of an expense item is passed through to apply against tax liability, under present U.S. federal and local tax laws, but capital items must be depreciated over a period of several years.

Given the present high cost and limited availability of capital to businesses, the future tax savings eventually obtained have a much reduced present value. A commonly expressed point of view is that the after-tax payback period should not exceed five years for business acceptability although some industries (notably utilities) may find longer periods acceptable. The discount rate applicable to business loans is typically higher, and the term shorter, than for residential mortgage financing. Offsetting this is the fact that the retail value of electricity to commercial customers is about 1.8 times greater than to residential customers under present pricing policies. Nevertheless, modification of the tax structure to permit accelerated depreciation or investment tax credits of about 50 percent (rather than the 10 percent presently applicable) will be needed before solar installations can be made with payback times in the 5 to 10 year range.

Relatively detailed studies have been made of hybrid photovoltaic/thermal total energy systems for business or residential installations given the climatic and insolation data for a number of U.S. cities including Boston, Phoenix, Fort Worth, Atlanta, Miami, Cleveland, Wilmington and others [1,2]. The conclusions differ in detail, but there is agreement that cost-effectiveness is approached more rapidly for residential than for business systems, for places like Phoenix and Fort Worth as compared to Boston, and for systems in which the collector mix is dominantly photovoltaic. Advanced heat-pump hybrid systems offer the best possibilities for saving, except in the colder locations where a simple solar hot water heat plus electric vapor-compression air conditioning system is most cost effective. This system should also be mostly photovoltaic in terms of collector mix, because of the higher electric prices prevailing in the Northern cities.

Depending on the validity of the specific assumptions of discount rate and escalation factor, and with lesser sensitivity, on the values of all other input variables as well, it appears that by 1985 residential hybrid systems providing 70 to 80 percent of total energy needs will be cost effective in many U.S. locations given an energy escalation rate of 3 percent. At a 6 percent escalation rate business installations in some locations (in the Southwest) become viable as well [1]. With somewhat different assumptions, only residential installations in the Southwest seem clearly cost effective for installation in 1985, but viability is reached in other areas well before 2000 [2].

There are other specialized intermediate term applications for photo-voltaic systems which are attractive at prices between those current and those of the 1986 DOE goals [3]. Crop irrigation provides a particular example with many advantages. Storage is built-in by the needs of the load, which are met by pumped water. Periods of cloud cover or pre-cipitation coincide with temporary reduction in the need for irrigation water; i.e., a crop may require a certain number of acre-feet of water per growing season but there is some flexibility in the necessity of pro-viding a particular quantity of water each day.

Photovoltaic irrigation systems may use electric storage to facilitate starting of pump motors, but frequency controlled inverters can also be used as an advantageous substitute [4]. Ac motors offer higher efficiency and reduced maintenance requirements, and are very much less expensive and more reliable than small ($\sim$10 hp) diesel engines which would provide competitive service. The electric output from the array can be used to power crop drying blowers and conveyor machinery at harvest periods when irrigation is not required. Several studies suggest that photovoltaic water pumping will be competitive with diesel powered pumps (or, in India [5], with animal-powered traditional methods) at module costs in the vicinity of $2.00/W_{pe}$, i.e., at the level of the 1982 DOE goals.

The largest demonstration project of this kind is in operation at the University of Nebraska Agricultural Research Station in Mead [6]. A 25 kW_{pe} array, a 90 kWhr lead-acid battery store, and power conditioning equipment appropriate for ac motor loads are included in the facility. The battery store is sized to provide smoothing to the load during sunlight hours, but *not* to provide power during cloudy or rainy days or at night. Several 5, 10 and 15 hp ac motor loads have been successfully started and operated and useful data on actual performance in irrigation and crop drying is being gathered.

Crop irrigation activities account for about 0.5 percent of the total energy consumption in the United States, so that as an intermediate or demonstration application of photovoltaic technology it is manageable but nontrivial in scope. In much of the Southwest, and in essentially all of the highly productive cropland of California, irrigation is a necessity; there the land has essentially no agricultural value without it. As conventional energy becomes more expensive, the value added by irrigation will be reduced and eventually become a negative factor, with a serious impact on the price and availability of food. An economic analysis of cropland irrigation in Arizona, Nebraska, Texas and California [7] suggests that under optimistic assumptions (a discount rate of 5 percent and escalation rate of 4 percent) the use of photovoltaic equipment for crop irrigation may become profitable as early as 1986 in Nebraska, and in 1983 in the other three states. With more conservative economic assumptions (8 percent discount rate and 2 percent escalation rate) the earliest time for profitable implementation moves out to $\sim$1990 in Arizona, California and Texas and 2000 in Nebraska. When the effect of present tax structure including the 10 percent investment tax credit is included, the optimistic dates for profitable installation move up about one year, while they occur several years earlier (1987 versus 1990 in California, for instance) for the conservative scenario. The analysis in this report predicts that relatively small additional tax credits would have a substantial effect in accelerating the introduction of photovoltaic technology. Since increased production and sales volume is expected to accelerate the downward trend in array prices, this change in public policy could have a major impact on the time scale for the general implementation of photovoltaic technology.

6.3 Social and Environmental Impact

The widespread manufacture and installation of photovoltaic systems will have an effect on patterns of employment, particularly in the energy producing sector of the economy. The recent trend there has been toward automation and reduction of the work force. While economic manufacture of photovoltaic devices will necessarily be highly automated, the sheer size of the manufacturing industry spawned will create many new jobs. Installation of systems will similarly have a significant impact on the construction industry, which will be of long-term duration since the transition to maximum utilization of photovoltaic power generation will take five to ten decades at least.

Photovoltaic generating systems are popularly believed to be totally clean and nonpolluting in operation. However, in addition to generation of pollution during operation, the environmental effects of manufacturing new systems and decommissioning or recycling old ones must be considered as well. In the case of electric generation by nuclear fission or fusion for example, the decommissioning problem is very serious indeed. Photovoltaic modules made from glass and silicon cells may be recycled back to sand or glass; elemental silicon is itself quite inert in the environment even thought it is not found naturally. The greatest danger during the traditional sliced-ingot Si cell manufacturing cycle is that posed by the escape of SiO_2 smoke and fumes during the initial quartzite-to-Si conversion. Inhalation of SiO_2 smoke leads to silicosis, a chronic lung disease and a potential hazard wherever SiO_2 is worked at high temperatures. Present industrial practice requires that glass blowing with silica be carried out only with adequate forced ventilation. On a large scale, of course, exhaust ventilation cannot be used since the worker's problem then becomes one for the general public.

A fair comparison of the emissions from an SiO_2 reduction plant would be to those from a coal-burning electric plant, since it appears that coal is the probable competitive energy source over the intermediate term. One should compare emissions associated with the same net production of energy, of course. A 100 MW coal fired plant will produce 10^9 kWhrs annually. One can take 3000 kWhrs per m^2 of array as the 20-year life output of 10 percent efficient modules. Hence the comparison Si plant must produce annually the equivalent of $4 \times 10^5 \, \text{m}^2$ of array area, or approximately 500 metric tons (after allowance for an 80 percent yield after kerf waste, edge trimming and all other losses in manufacturing). The production of Si from quartzite will be similar insofar as emission problems are concerned to present metallurgical Si production, for which emissions data can be obtained. There appears to be no question that coal fired electric generators, even with the best possible scrubbing technology now available, will be a much more serious source of air pollution [8].

Other materials of a less stable and benign nature are used in the fabrication of silicon cells and modules. The volatile compounds of boron and phosphorous used to form device junctions are highly toxic, but they are incorporated at very low levels (one part per million or less in the actual devices) and the low volume of use permits complete containment at the manufacturing facility. Various adhesives, plastics and potting resins are used as well, but these will not represent large additions to the present plastic waste burden. The use of freons or other halogenated organic solvents in surface cleaning steps will probably be necessary. This is a matter of concern both for the possible buildup of heavier chlorinated organics in the ground water and river systems and for the effect of the lighter, volatile compounds on the ozone layer in the earth's atmosphere. Techniques to ensure containment of these materials at the manufacturing site must be developed, but the overall effect would appear to be negligible in comparison to the present Freon burden from automobile air conditioners and existing sources of halogenated solvent emission, e.g., the dry-cleaning industry.

Other potential solar cell materials do not present such an optimistic picture, either from the point of view of worker health and safety on the job or from the point of view of exposure of the general public to low levels of toxic materials. Cd is a highly toxic element which tends to be accumulated in the body, leading to kidney and liver disorders. It has a relatively high vapor pressure as do the compounds it would be expected to form in the environment. Fire damage to CdS/Cu_2S modules would result in emission of volatile Cd oxides, of much more concern than the oxides of Cu and S released simultaneously. CdS hydrolyses slowly to form soluble oxides of Cd, so that weathering of broken modules would also release Cd into the environment. During manufacture the Cd is necessarily confined within closed vacuum systems, and manufacture would seem to pose much less of a problem than post-installation accident. Recycling used or defective modules will be necessary since Cd is 100 times less abundant than Ga (10^6 times less abundant than Si). Experience with the present extraction techniques by which Cd is obtained from ores should be adequate for this, since used solar modules would, of course, represent very high grade "ore deposits" of the constituent elements.

The other material system under present consideration in practical contexts is based on GaAs. In LPE growth containment of the reactants is straightforward and explosion problems associated with the storage and use of hydrogen are the primary concern. Arsenic and its compounds have been known since early historic times to be poisonous. Inorganic arsenic compounds have recently been designated as carcinogenic as well. The essence of current legislation in the United States is that there is *no* acceptable maximum level for known carcinogens in the environment. Large scale manufacture of a product consisting over half by weight of such an agent obviously will be fraught with problems. Fire and hydrolytic weathering release water soluble oxides of arsenic, and hence damage to arrays of GaAs cells will present a potentially serious environmental problem.

Other manufacturing techniques which could be used for GaAs cell production are less tractable than LPE from the point of view of industrial and environmental safety. The metallorganic compounds of Ga and Al are volatile, explosive, and pyrophoric; which makes on-site safety a concern, but minimizes the problem with regard to long term effects in the environment. Al_2O_3 and, so far as is known, Ga_2O_3 as well, are innocuous (although one could anticipate problems akin to silicosis if finely divided aerosols were inhaled).

Arsine is an extremely toxic compound which is absorbed directly into the blood on inhalation. It is incorporated intact into hemoglobin and subsequent rupture (hemolysis) of the red blood cells results. This frees the arsine-hemoglobin complex, which is taken up by another red blood cell, producing a quasi-catalytic destruction of these bodies. The probability for oxidation of the arsine molecule to the relatively less toxic oxide form is only 10^{-2} to 10^{-3}, and no effective antidote is known. The body's natural capacity to make new red blood cells allows a steady atmospheric background level of only a few tens of parts per billion of AsH_3 to be tolerated for continued 8-hour exposure.

The metallorganic growth process is only about 10 percent efficient in converting the As in the input arsine into the desired GaAs product. Hence a large quantity of exhausted arsenic in elemental or oxide form must be processed, and a large volume of arsine must be handled (and manufactured in the first place). Fortunately arsine is unstable in the environment and is rapidly decomposed to oxide forms in contact with water and soil.

Phosphide compounds also present problems. InP weathers to form PH_3 under some conditions, which while not so insidious a poison as AsH_3 is still highly toxic. It is also more stable and can be converted in the environment into organophosphorus forms rather than simply oxidized to inert phosphates. Phosphorous is taken up readily by life systems, usually with acute toxic effect on the nervous system and chronic effect on the kidneys and liver. In the manufacturing processes, conversion of input PH_3 to elemental phosphorous often results in a mixture of allotropic forms containing some white phosphorous (P_2) which is pyrophoric and ignites the remainder upon exposure to air. Cleaning exhaust traps and lines is therefore rendered difficult, and the likelihood of release of phosphorous oxide smoke is high.

The term "environmental and social cost" may certainly be construed to include the visual impact of large areas of photovoltaic collectors in populated areas. The total area in question $(\sim 2 \times 10^9 m^2)$ is of the same order of magnitude as that now occupied in the United States by roof-tops of building structures, or that converted *each year* to asphalt and concrete paving. It seems likely that architectural tradition will change and de-centralized photovoltaic systems will be accepted. There is no question of having to cover the entire landscape with photovoltaic arrays, and there seems to be little unique architectural merit in the majority of present rooftops which would be lost in any case.

The impact of extensive centralized photovoltaic generation would be significantly different and more intrusive. A 100 MW plant would require $6 \times 10^6 m^2$ of collector area, covering a square two miles on a side. Large areas of land near urban and suburban load centers would thus be foreclosed from other possible uses. The presence of high voltage and high current collectors would require fencing or other isolation of the area from the public. Use of the area under the arrays for business or recreational pursuits would not seem to be practical.

An alternative central generation concept utilizing an orbiting solar array was initially proposed over a decade ago [9] and has recently aroused substantial renewed interest [10]. In stationary earth orbit (at a height of 22,000 miles) a satellite is illuminated nearly all the time, with the earth eclipsing the sun for only a few tens of hours per year. The proposal involves the conversion of the direct current photovoltaic power to microwave radiation which is to be beamed to receiving antenna grids on the earth's surface. The principal advantage of this scheme is that power is available on a 24-hour basis. There are obvious problems associated with the satellite construction and maintenance in orbit. The cost has been estimated at $1700/kW (as compared to $1400/kW for nuclear plants) assuming an array of 50 satellites each 25 × 5 km in size and producing 10 GW. An approximate total program cost of 500 billion dollars has been stated. The value of electricity produced over a 30-year lifetime would be about 3.5 trillion dollars (at the wholesale rate of $.027/kWhr), however, so the scheme should be cost effective [11]. An average retail cost of solar-satellite derived electricity of \$0.05 to \$0.07 per kWhr has been estimated for the 2010—2020 time period [12].

Each satellite would beam microwave power to a $5 \times 10^7 m^2$ rectenna grid. In addition to the environmental impact of this grid, the possible effect of the microwave beams on aircraft and living creatures is a matter of concern. The low power density (0.02 W/cm^2) is not believed to pose a health problem for short-term exposure, but the possibility that genetic damage (and subsequent birth defects) might result from low-level exposure of a large population may require further investigation.

The cost estimates for this "Sun-Sat" program are highly uncertain and depend heavily on the successful development of advanced, reusable launch vehicles. The requisite photovoltaic and microwave technology is, however, well established. The cost of the solar cells is less important than for terrestrial systems and emphasis would be placed on performance and operating lifetime. Provided the microwave radiation levels are indeed found to be harmless, this most far-fetched sounding of the photovoltaic proposals seems technically feasible, particularly when contrasted to other futuristic schemes such as the as yet undemonstrated generation of electricity from controlled nuclear fusion.

6.4 Solar Energy and Inflation

In the United States, and indeed in most of the world today, inflation and erosion of the value of the national currency is a pressing problem. Inflation breeds social unrest in developing as well as in industrially developed countries. A reality throughout the world is that energy costs more today than in the recent past. It is important to realize that, although sunlight is indeed free, solar energy, and in particular solar photovoltaic electric power, is not only *not* free, it will be expensive in the future even in comparison to today's inflated energy prices. It is also important, however, to recognize that the fact that the solar fuel *is* free makes solar energy utilization uniquely anti-inflationary.

With the exception of hydroelectric energy (which is susceptible to only limited further development in most countries and is, in any case, a form of solar energy) and the as yet unproved development of fusion energy, other practical energy options are predicated upon accelerated depletion of nonrenewable mineral resources. In the United States, for example, increased exploitation of natural gas and coal will lead to rapidly rising prices and further inflation; not only because new wells and new mines use new equipment and materials (at today's, not yesterday's prices), but because they must be drilled deeper or process greater volumes of low grade deposit for the same yield. There is no such increase in the price of sunlight, regardless of the degree of utilization of solar photovoltaic or thermal collectors, or the rapidity of their introduction.

Purchase of fossil fuels from foreign sources has a directly inflationary effect as those foreign fuel exporters seek to redeem their currency surplus by purchasing consumer goods from the debtor nation. This additional demand for consumer goods and the excess supply of currency returning from abroad combine to drive up prices of consumer items sharply. In contrast the expenditure of funds on solar energy equipment effectively *removes* currency from circulation, and *reduces* demand for consumer goods. This diversion of resources from consumption to investment takes place equally effectively whether the photovoltaic equipment is owned by private individuals or utilities (who divert the cash from customers through higher rates), but a psychological advantage may exist if individuals own the equipment. They are likely to feel that they own a capital asset, which they can later resell, and are less likely to press for higher wages or income to "get even" with increased energy expense than if their resources are externally diverted by increased utility bills or taxes.

The production of solar thermal collectors has only recently become a commercial reality, and the production of photovoltaic modules will not be commercialized for several years at best. Nevertheless, the more solar energy is used, the cheaper it will become. The production of solar thermal modules will benefit from economies of scale and automated, large volume manufacture, but will create primarily low-technology jobs for semiskilled workers and construction laborers. Photovoltaic production will require and benefit from engineering improvements, and hence will provide high technology employment for skilled workers. However, many of the assembly and installation tasks associated with photovoltaic systems require only a level of skill comparable to that necessary for the installation of solar thermal equipment. The inherent, long range limiting cost of photovoltaic and solar thermal modules is similar. Thermal efficiencies are higher, but the value of electric energy is greater than the value of heat by an approximately offsetting factor. We may expect the price of either thermal or electric energy from solar systems to stabilize at two to three times the price of today's fossil-fuel energy — a price level which will be reached well before the end of this century if present trends continue.

As solar energy displaces gas and oil, the reduced demands for those fuels will be reflected in even further anti-inflationary trends as the price of petrochemical feedstocks and motor lubricants stabilizes, or at least climbs less rapidly. A super-linear reduction in the rate of consumer price increases can thus be expected to result from the displacement of nonrenewable fuels by solar energy.

Government spending, whether in the guise of tax credits or direct subsidy, must be offset by deficit financing or increased tax revenues from other sources, and will inevitably have some inflationary effect. Tax credits given for installed solar energy equipment are paid only *after* the equipment is in place, saving fuel, and hence contributing an offsetting, anti-inflationary effect. The continued granting of subsidies for accelerated exploitation of finite domestic fuel resources, however, is like the sowing of dragons teeth, having the same super-inflationary effect as expanded purchase of foreign fuels. The deflationary effect of locking up capital in solar arrays will be offset by the growth of a new solar energy industry and the jobs it creates. The manufacture of photovoltaic modules in the U.S. alone would spawn an industry comparable in size to the automotive industry if the DOE program goals are to be materialized.

In the intermediate term developed nations have no practical alternative to the continued consumption of fossil and fissile fuels. It is unrealistic to expect a breakthrough in photovoltaic technology which will result either in cheap electric energy or a more rapid implementation of photovoltaic systems than suggested by the DOE program goals. In the long term, large scale photovoltaic conversion, probably based on crystalline or semicrystalline silicon cells, will provide a substantial portion of the world's energy needs, without contributing thermal, chemical or radioactive pollution to the biosphere. Many difficult engineering problems must be overcome, but these appear to involve questions of energy storage, module encapsulation, and production line automation more than questions pertinent to the photovoltaic cells proper. The day of fuels mined from the earth is drawing to a close, and the age of solar energy is dawning. Solar cell arrays will be as important a part of the future as gasoline and steam engines have been in the present and recent past.

REFERENCES

1. E. C. Kern, Jr. and M. C. Russell, *Hybrid Photovoltaic / Thermal Solar Energy Systems, Report C00-4577-1*, MIT Lincoln Laboratory (1978, unpubl.).

2. E. F. Federman, W. Feduska, W. J. McAllister and S. L. Nearhoof, *Proc. 13 PVSC,* 1004 (IEEE, New York, NY, 1978).

3. P. F. Pittman and C. R. Chowaniec, *Proc. 13 PVSC,* 1010 (IEEE, New York, NY, 1978).

4. E. E. Landsman, *Proc. 13 PVSC,* 992 (IEEE, New York, NY, 1978).

5. S. K. Sangal, Y. A. Gnanainder and E. V. Bhaskar, *Proc. 13 PVSC,* 1278 (IEEE, New York, NY, 1978).

6. L. L. Bucciarelli, J. D. Cremin, A. A. Fenton, R. F. Hopkinson, E. F. Lyon and W. R. Romaine, *Proc. 13 PVSC,* 16 (IEEE, New York, NY, 1978).

7. R. W. Matlin and M. T. Katzman *The Economics of Adopting Solar Photovoltaic Energy Systems in Irrigation, Report C00/4094-2,* MIT Lincoln Laboratory (1977, unpubl.).

8. T. L. Neff, *Proc. 13 PVSC,* 1001 (IEEE, New York, NY, 1978).

9. P. E. Glaser, *Science* **162,** 857 (1968).

10. cf. *Industrial Research/Development,* July, 1978 (p. 35).

11. cf. *Solar Power from Satellites — Hearing Record of the Senate Subcommittee on Aerospace Technology and National Needs, 94th Congress (Jan, 19 and 21, 1976).* U.S. Government Printing Office, Washington, DC (1976).

12. G. R. Woodcock, *J. Energy* **2,** 196 (1978).

Bibliography

In addition to the technical sources particularly cited in the text, the reader is referred to the following four books which provide different depths of treatment and/or emphasize different aspects of photovoltaic science and technology.

Backus, C. E., Ed. *Solar Cells.* IEEE Press, New York, NY, 1976. A compilation of classic papers in the photovoltaics field which originally appeared in the 1954 to 1974 period.

Hovel, H. J. *Solar Cells, Adv. in Semiconductors and Semimetals, vol. XI.* R. K. Willardson and A. C. Beer, Eds. Academic Press, New York, NY, 1975. This contains a detailed treatment of the physics and technology of single crystal silicon solar cells.

Pulfrey, D. L. *Photovoltaic Power Generation.* Reinhold, New York, NY, 1978.

Palz, W. *Solar Electricity — An Economic Approach to Solar Energy.* UNESCO — Butterworths, London, UK and Boston, MA, 1978.

In January of 1979 the American Physical Society Study Group on Solar Photovoltaic Conversion made available a preprint version of their *Principal Conclusions,* a more extensive presentation of which is expected to appear in 1980 in *Reviews of Modern Physics.* The reader may wish to compare this less engineering-oriented view to that which has been presented here.

Index